Policy, Practice and Education of the Circular Economy in Built Environments

nova
science publishers
www.novapublishers.com

Waste and Waste Management

More information about this series can be found at
https://novapublishers.com/product-category/series/waste-and-waste-management/

Construction Materials and Engineering

More information about this series can be found at
https://novapublishers.com/product-category/series/construction-materials-and-engineering/

Salman Shooshtarian
Peter SP Wong
Tayyab Maqsood
Editors

Policy, Practice and Education of the Circular Economy in Built Environments

nova
science publishers
www.novapublishers.com

DOI: https://doi.org/10.52305/URCG0650

Library of Congress Cataloging-in-Publication Data

ISBN: 979-8-89530-623-9 (Hardcover)
ISBN: 979-8-89530-931-5 (eBook)

Published by Nova Science Publishers, Inc. † New York

Contents

Acknowledgments

This book is based on a research project that has been developed by Australia's Sustainable Built Environment National Research Centre (SBEnrc). The SBEnrc develops projects informed by industry partner needs, secures national funding, project manages collaborative research and oversees research into practice initiatives. Current core Members of the SBEnrc include ATCO Australia, BGC Australia, Government of Western Australia, Queensland Government, Transport for NSW, Sunshine Coast Council, Griffith University, Western Sydney University, Curtin University and RMIT University. The industry-driven research outlined in this publication would not have been possible without the valuable support of our core industry, government and research partners.

List of Abbreviations

3R	Reduction, Reuse, Recycling
A&D	Architecture and Design
ABS	Australian Bureau of Statistics
ACCC	Australian Competition and Consumer Commission
ACT	Australian Capital Territory
AGF	Asphalt Granite Fines
APC	Australian Packaging Covenant
APCC	Australasian Procurement and Construction Council
C&D	Construction and Demolition
C&I	Commercial and Industrial
CBM	Circular Business Model
CBSM	Community-Based Social Marketing
CCI	Circular Construction Indicator
CCS	Cluster Corporate Services
CDS	Container Deposit Scheme
CE	Circular Economy
CEIP	Circular Economy Indicator Prototype
CI	Material-based Circularity Index
CPI	Circularity Potential Indicator
CSCM	Circular Supply Chain Management
CSIRO	Commonwealth Scientific and Industrial Research Organisation
CSR	Corporate Social Responsibility
CTI	Circular Transition Indicators
DCEEW	Department of Climate Change, Energy, the Environment and Water
DFAT	Department of Foreign Affairs and Trade
DfD	Design for Deconstruction
DfD	Design for Disassembly
DGNB	Deutsche Gesellschaft für Nachhaltiges Bauen

DJSIR	Department of Jobs, Skills, Industry and Regions
DWER	Department of Water and Environmental Regulation
EfW	Energy from Waste
EMF	Ellen McArthur Foundation
EoL	End of Life
EOL-RIR	End-of-Life Recycling Input Rate
EoW	End of Waste
EPA	Environmental Protection Authority
EPD	Environmental Product Declaration
EPR	Extended Producer Responsibility
ESG	Environmental, Social and Governance
ETPB	Extended Theory of Planned Behaviour
ETS	Emission Trading System
EU	European Union
FSC	Forest Stewardship Council
GBCA	Green Building Council of Australia
GBI	Green Building Index
GDP	Gross Domestic Product
GECA	Good Environmental Choice Australia
GGT	Global GreenTag
GHG	Greenhouse Gas Emissions
GISA	Green Industries SA
GREP	NSW Government Resource Efficiency Policy
GRI	Global Reporting Initiative
HSEQ	Health, Safety, Environment and Quality
ILFI	International Living Future Institute
IPWEA	Institute of Public Works Engineering Australasia
ISC	Infrastructure Sustainability Council
ISL	International Synergies
ISO	International Organization for Standardisation
KPI	Key Performance Indicator
LCA	Life Cycle Analysis
LEED	Leadership in Energy and Environmental Design
LGA	Local Government Area
LGNSW	Local Government NSW
m	Million
M-CEFBC	Material Circularity Evaluation Framework for Building Construction
MCI	Material Circularity Indicator

MMC	Modern Methods of Construction
MR	Material Reutilisation
MRF	Material Recycling Facility
MSW	Municipal Solid Waste
Mt	Million tonnes
NABERS	National Australian Built Environment Rating System
NAT	Norm Activation Theory
NCC	National Construction Code
NCI	New Circularity Index
NGER	National Greenhouse and Energy Reporting
NISP	National Industrial Symbiosis Programme
NSW	New South Wales
NT	Northern Territory
NTCRS	National Television and Computer Recycling Scheme
NWP	National Waste Policy
OECD	Organisation for Economic Co-operation and Development
OEH	Office of Environment and Heritage
PC	Product Circularity
PEFC	Programme for the Endorsement of Forest Certification
PLCM	Product-Level Circularity Metric
PMO	Project Management Office
PP	Proximity Principle
PPP	Public Private Partnership
PRC	Recycled Product Certification
PRISMA	Preferred Reporting Items for Systematic Reviews and Meta-Analyses
PVC	Polyvinyl Chloride
PwRC	Product with Recycled Content
QLD	Queensland
R&D	Research and Development
RAP	Reclaimed Asphalt Pavement
RCA	Recycled Concrete Aggregate
RLR	Rapid Literature Review
ROI	Return on Investment
RPC	Recycled Product Certification
RPI	Reuse Potential Indicator
RtR	Roads to Reuse
SA	South Australia
SBEnrc	Sustainable Built Environment National Research Centre

SDG	Sustainable Development Goal
SIT	Social Interaction Theory
SLR	Systematic Literature Review
SMEs	Small and Medium Enterprises
SRLA	Suburban Rail Loop Authority
SV	Sustainability Victoria
Tas	Tasmania
TPB	Theory of Planned Behaviour
UAE	United Arab Emirates
UDIA	Urban Development Institute of Australia
UK	United Kingdom
UN	United Nations
USGBC	US Green Building Council
Vic	Victoria
VIDA	Victorian Infrastructure Delivery Authority
WA	Western Australia
WMRR	Waste Management and Resource Recovery Association of Australia
WRAP	Waste & Resources Action Program.

Chapter 1

Introduction

Salman Shooshtarian
School of Property, Construction and Project Management, RMIT University, Melbourne, Australia

Abstract

Adopting a circular economy model has become essential to advancing sustainability across many sectors, notably in building and construction. At its core, this approach envisions an industrial system that is inherently restorative and regenerative. It offers a range of strategies designed to minimise landfill waste and lower carbon emissions. Yet, integrating circular economy practices into the construction industry is far from straightforward. Challenges arise from conflicting stakeholder interests, financial and regulatory hurdles, and various environmental concerns. In response, a host of initiatives - both local and global - have emerged to effectively manage construction and demolition (C&D) waste materials. These initiatives are typically classified into enforcement, encouragement, and education approaches. This chapter outlines these three domains and establishes the link between them and the following chapters. This chapter - and the book as a whole - emphasises the integration of recycled-content products (PwRC) into construction projects. This integration is among the top current priorities in applying circular economy principles in the building and construction sector.

Keywords: circular economy, education, enforcement, encouragement, products with recycled content

In: Policy, Practice and Education of Circular Economy in Built Environments
Editors: Salman Shooshtarian, Peter SP Wong and Tayyab Maqsood
ISBN: 979-8-89530-623-9

1. Introduction

The building and construction sector is among the sectors that are known for its low resource efficiency. Globally, the sector is responsible for generating substantial quantities of solid waste and carbon emissions [1]. Traditional construction practices have largely adhered to a linear model: take-make-dispose of. This model follows a process wherein raw materials are extracted, processed into construction components, used during a building's lifecycle, and ultimately disposed of, resulting in the irreversible loss of embedded material value. The sheer volume of waste produced, often ending up in landfills [2], coupled with the industry's heavy reliance on finite natural resources [3], necessitates a fundamental shift in practices.

Globally, numerous initiatives and policies are emerging to guide industry and government agencies in achieving resource efficiency, waste minimisation, net-zero targets, and the United Nations Sustainable Development Goals (SDGs). The circular economy is an increasingly popular concept that promises to solve the aforementioned issues. The Ellen MacArthur Foundation (EMF) articulates the circular economy as an industrial system that is "restorative or regenerative by design" [4]. This definition encapsulates a transformative shift away from the conventional linear "take-make-dispose" model towards a systemic approach that emphasises the perpetual retention of material value. In the circular economy framework, waste is not an inevitable by-product of production; rather, it is regarded as a resource to be reintegrated into the manufacturing cycle. Materials, energy, and products are continuously cycled back into the economy, thereby reducing the need for new resource extraction and mitigating environmental impacts.

The most recent Circularity Gap Report 2024 states that our current economy is only 7.2% circular, leaving space for CE policies that [5], combined with the climate change agenda, can foster a path towards sustainable use of resources globally [6]. Australia, like other Organisation for Economic Co-operation and Development (OECD) nations, is undergoing a significant transition towards a circular economy. In recent years, Australia has introduced several exemplary policies and initiatives, including APCO Packaging Stewardship, Nature Positive policy [7], Environmental, Social and Governance (ESG) reporting, and Commonwealth Climate Disclosure. Hence, Australian organisations across various sectors are increasingly adopting circular economy business models to align with local, national, and international circular economy trends. In December 2024, the Australian

Federal Department of Climate Change, Energy and the Environment and Water (DCCEEW) published the first national Circular Economy Framework [8]. This framework highlights the need to double the country's circularity rate by 2035. The modelling study by the Commonwealth Scientific and Industrial Research Organisation (CSIRO) indicates that within the next decade committing to the ambitious goal of this framework can result in several positive outcomes [9]. These include adding an annual $26 billion to the country's Gross Domestic Product (GDP), a reduction in greenhouse gas emissions by 14%, and an annual diversion of 26 million tonnes of materials from landfills.

In Australia's Circular Economy Framework, the built environment is emphasised as a priority sector, accounting for one-third of global resource consumption [10] and serving as the principal destination for materials in Australia [11]. Alongside refurbishing and adaptive reuse, prioritising the use of sustainable construction materials, including products with recycled content (PwRC), is strongly encouraged to drive the sector's shift towards a circular economy. This strategy is gaining traction and is being adopted in several construction projects nationwide [12]. The framework proposes that the optimal utilisation of these resources can be achieved by fostering PwRC markets. However, creating demand for PwRC is not easy, as it is affected by various internal and external factors that the editors of this book identified in their earlier research [13].

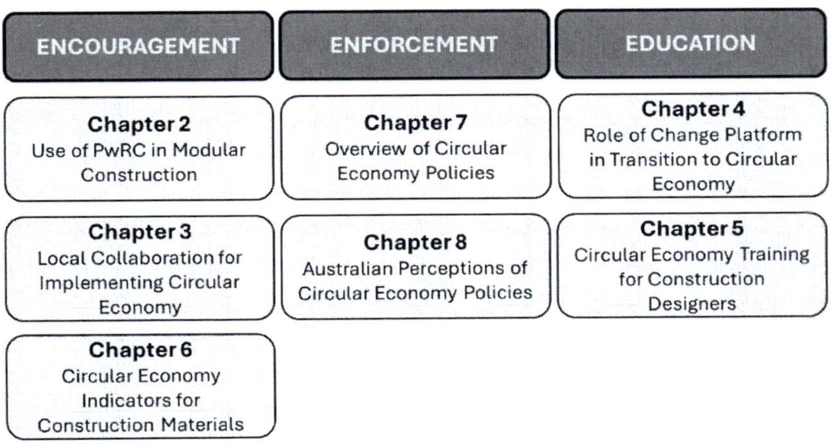

Source: Author.

Figure 1. An overview of the book's structure that aligns with the approaches of encouragement, enforcement, and education.

Most initiatives designed to promote a circular economy through optimizing the use of PwRC are typically categorized under one or more of three approaches: encouragement, enhancement, and enforcement. It is widely believed that successfully planning a transition to a circular economy requires the simultaneous application of all three approaches. As illustrated in Figure 1, this book aims to provide a clear insight into how these approaches can effectively foster circularity in the building and construction sector. The following sections briefly examine the three approaches outlined above, highlighting how they are integrated within the book's chapters.

2. Encouragement

The building and construction sector has long been characterized by its conservative approach to integrating new technologies and concepts - a trend that is even more pronounced when it comes to sustainability. Earlier studies [14, 15] have consistently shown that the sector's uptake of sustainable technologies and practices lags considerably behind that of other industries. This slow adoption can be attributed to several factors, including the reliance on established traditional methods [16], regulatory and financial constraints, and the inherently complex nature of construction projects [12]. As a result, there is a pressing need for innovative strategies and targeted interventions to encourage the adoption of sustainable practices including circular economy principles.

Encouraging the sector to adopt circular economy principles- including the use of PwRC- is primarily driven by three key factors: financial return, social image and government supportive policies. Financial incentives can translate into significant cost savings, enhanced efficiency, and new revenue opportunities. At the same time, a strong social image helps organisations build trust and credibility with stakeholders, positioning them as leaders in sustainability. A recent study in Australia [12] reports that 'social image and recognition for environmental sustainability' is among the top five factors encouraging the use of PwRC in the studied construction projects.

Meanwhile, proactive government policies create an enabling environment by offering incentives, reducing regulatory hurdles, and setting industry standards [12]. Baah, Jin [17] noted that regulatory authorities can exercise their power to encourage organisations to adopt green practices, which translates into higher market share and profitability and finally grants firms some social legitimacy. According to Aryee and Kanda [18], experts argue that responding to regulatory pressures enhances the organisations' positive image or social standing. Together, these factors provide a compelling business case for the sector to transition toward more sustainable, circular practices.

Drawing on three case studies, Chapters 2, 3, and 6 delve into three targeted encouragement strategies that have the potential to enhance circularity in construction projects. Chapter 2 examines the role of modular construction and local collaborations in optimising the use of PwRC, while Chapter 6 focuses on establishing circular economy indicators for these resources. Together, these chapters illustrate how embracing new PwRC applications can motivate and empower stakeholders to drive a more sustainable and circular approach in the construction sector.

3. Enforcement

An interesting aspect of Australia's Circular Economy Framework [8] is its inclusive approach, which involves not just the government but also industry, investors, communities, NGOs and academics. It stresses that managing waste and using resources sustainably is a shared responsibility, which is key to making the circular economy work. The framework specifically highlights the government's role in creating supportive policies:

> 'Policymakers set the direction and provide the foundation for a successful transition to a circular economy. Governments can create the conditions that allow circular systems to thrive by implementing supportive policies, providing financial incentives, and encouraging collaboration and behaviour change across sectors' [8, p. 30].

At the international level, growing evidence has underscored the crucial role of developing circular economy policies in promoting the use of PwRC within the sector. Nußholz, Rasmussen [19] proposed that public policies can support organisations to address circular economy barriers by incorporating higher recovery goals and incentivising the PwRC market. In China, Bao and Lu [20] pointed out that strong governmental interventions combining penalties and subsidies are an important factor contributing to successful circular economy implementation. In the Nordic region, a study by zu Castell-Rüdenhausen, Wahlström [21] indicates that many business opportunities in the sector are supported by national and local policies. Consequently, translating circular economy principles into actionable policies can indeed drive the building and construction sector's transition toward a circular economy.

It should be acknowledged that not all circular economy policies necessarily yield positive results for optimal PwRC utilisation. Existing research shows that, depending on various internal and external factors, circular economy policies can produce either beneficial or detrimental outcomes [22-24]. For instance, Shooshtarian, Maqsood [22] contend that while landfill levies may be the preferred policy approach in certain scenarios, they can also inadvertently discourage best waste management practices. These levies may lead to increased illegal dumping, shifting of waste to other regions, and a reduction in overall recycling efforts. Rameezdeen, Chileshe [25] show that illegal dumping can be an unintended consequence of a higher landfill levy rate. Hence, it is crucial to assess the unintended consequences associated with developing and implementing circular economy policies within the context of the sector. The most effective way to gain this understanding is through systematically gathering insights and perceptions from both industry and government stakeholders.

Chapters 7 and 8 in this book provide a comprehensive exploration of circular economy policies that can directly or indirectly influence the use of PwRC in the building and construction sector. Chapter 7 presents an in-depth overview of 17 policies, each documented for its role in facilitating the transition to a circular economy in construction projects. In Chapter 8, findings from a national survey are explored which offers insights into the effectiveness and characteristics of these policies.

3. Education

Education is pivotal in paving the way toward a circular economy. Any transformative change in waste management systems and sustainability practices must begin by empowering key stakeholders with the right knowledge and skills [26]. In recent years, organisations around the globe have launched industry-driven training programs designed to cultivate a circular mindset, ensuring that every participant is equipped to contribute to a more sustainable future.

Several studies have highlighted the critical role of education in the circular built environment context. Results of a survey showed that 'culture, poor education (attitude and behaviour) and acceptance' are among the top four factors hindering a successful C&D waste management system [27]. In another research focusing on stakeholders knowledge of circular economy, it was found that close to 70% of research participants indicated that they understood the meaning of the circular economy concept; however, only about 13% of them had adequate knowledge to adopt and apply it in their businesses and organisations [28]. Tam [29] argued that inadequate private sector education on waste management technology investment is a key barrier to the adoption of C&D waste-related technologies in the sector.

There is an ongoing debate about the individual role of education in promoting sustainable behaviour and attitudes in this context. For instance, Crocker and Lehmann [30] highlight that when it comes to sustainable behaviour, education alone has little or no effect, concluding that improving knowledge and changing attitudes may not impact behaviour. Furthermore, a study in Australia [31] found that decisions on circular economy practices in the sector are based on their financial return. The authors highlight the importance of imposing legislation to enhance circular economy practices until such practices become culturally embedded in organisations across the supply chain. Therefore, it can be inferred that education and an associated cultural shift need to be aligned with encouragement and enforcement approaches. On the other hand, Chapman, Skinner [32] observed that experiential learning results in new habits, and actively experiencing things is the most effective way to learn about environmental initiatives.

In Australia, the National Waste Policy [33] Strategy 3 (knowledge sharing, education and behaviour change), indicates that coordinated knowledge-sharing and education initiatives should be implemented to address the needs of governments, businesses and individuals, and to encourage the redesign, reuse, repair, resource recovery, recycling and reprocessing of products. Most Australian jurisdictions have formed teams dedicated to educating industry people and the wider community. Research and education are critical factors highlighted in all Australian waste strategy and guideline documents. These documents have introduced initiatives to fund education programs to reduce waste disposal [34]. A report by the Senate Environment and Communications References Committee [35] recommended that the Australian Government should support state and local governments in providing effective recycling education initiatives.

Achieving a circular built environment through education requires robust collaboration among diverse stakeholders. Educators, industry professionals, policymakers, and community members must come together to ensure that educational initiatives are not only relevant and effective but also truly transformative. In Chapter 5, this book delves into the critical role of training in driving the transition toward a circular economy within the built environment sector. This chapter is especially relevant for the architecture and design (A&D) community, which is uniquely positioned to implement a variety of innovative circular design strategies [36]. The chapter showcases the journey of a team of educators as they develop and deliver a training package aimed at cultivating circular thinking among A&D professionals in Australia.

References

[1] EuRIC, *Increasing recycling in the EU construction sector*. 2023, European Recycling Industries' Confederation.

[2] Ali, A. K., Architecture, Waste, and the Circular Economy. *Enquiry The ARCC Journal for Architectural Research*, 2021. 18(2): p. 1-3.

[3] Santos, P., et al., Circular Material Usage Strategies and Principles in Buildings: A Review. *Buildings*, 2024. 14(1): p. 281.

[4] Ellen MacArthur Foundation, *What is a circular economy?* . 2019.

[5] Circle Economy, *The Circularity Gap Report 2024*. 2024, Circle Economy.

[6] Fraser, M., Haigh, L. and Soria, A. C., The Circularity Gap Report 2023. 2023.

[7] DCCEEW, *Nature Positive Plan: better for the environment, better for business*, E. Department of Climate Change, the Environment and Water, Editor. 2022: Canberra, Australia.

[8] DCCEEW, *Australia's Circular Economy Framework: Doubling our circularity rate*, E. Department of Climate Change, the Environment and Water, Editor. 2024, Australian Government: Canberra, Australia.

[9] CSIRO, *Modelling circular economy transition targets*, A.s.N.S. Agency, Editor. 2024, CSIRO: Melbourne, Australia.

[10] Carpentier, H., *Circular Economy in the built environment waste hierarchy: Why recycling is the last resort*, in *Thought Leadership*. 2023, World Green Building Council.

[11] Encycle Consulting, *A study into commercial & industrial (C&I) waste and recycling in Australia by industry division*. 2013, Department of Climate Change, Energy, the Environment and Water: Perth, Australia.

[12] Shooshtarian, S., Maqsood, T., Wong, P. S., Caldera, S., Ryley, T., Zaman, A. and Ruiz, A., Circular economy in action: The application of products with recycled content in construction projects: A multiple case study approach. *Smart and Sustainable Built Environment*, 2024. 13(2): p. 370-394.

[13] Shooshtarian, S., Caldera, H., Maqsood, T., Ryley, T., Wong, S. P. and Zaman, A., Analysis of factors influencing the creation and stimulation of the Australian market for recycled construction and demolition waste products. *Sustainable Production and Consumption*, 2022. 34(2022): p. 163-176.

[14] Dadzie, J., Runeson, G., Ding, G. and Bondinuba, F. K., Barriers to Adoption of Sustainable Technologies for Energy-Efficient Building Upgrade—Semi-Structured Interviews. *Buildings*, 2018. 8(4): p. 57.

[15] Hofman, B., de Vries, G. and van de Kaa, G., Keeping things as they are: How status quo biases and traditions along with a lack of information transparency in the building industry slow down the adoption of innovative sustainable technologies. *Sustainability*, 2022. 14(13): p. 8188.

[16] Kineber, A. F., et al., Modelling the relationship between digital twins implementation barriers and sustainability pillars: Insights from building and construction sector. *Sustainable Cities and Society*, 2023. 99: p. 104930.

[17] Baah, C., Jin, Z. and Tang, L., Organizational and regulatory stakeholder pressures friends or foes to green logistics practices and financial performance: Investigating corporate reputation as a missing link. *Journal of Cleaner Production*, 2020. 247: p. 119125.

[18] Aryee, R. and Kanda, W., A strategic framework for analysing the effects of circular economy practices on firm performance. *Journal of Cleaner Production*, 2024. 476: p. 143753.

[19] Nußholz, J. L., Rasmussen, F. N., Whalen, K. and Plepys, A., Material reuse in buildings: Implications of a circular business model for sustainable value creation. *Journal of Cleaner Production*, 2020. 245: p. 118546.

[20] Bao, Z. and Lu, W., Developing efficient circularity for construction and demolition waste management in fast emerging economies: Lessons learned from Shenzhen, China. *Science of The Total Environment*, 2020. 724: p. 138264.

[21] zu Castell-Rüdenhausen, M., Wahlström, M., Astrup, T. F., Jensen, C., Oberender, A., Johansson, P. and Waerner, E. R., Policies as drivers for circular economy in the construction sector in the Nordics. *Sustainability*, 2021. 13(16): p. 9350.

[22] Shooshtarian, S., Maqsood, T., Khalfan, M., Yang, J. R. and Wong, S. P. P., Landfill levy imposition on construction and demolition waste: Australian stakeholders' perceptions *Sustainability*, 2020. 12(11): p. 4496.

[23] Yu, S., Awasthi, A. K., Ma, W., Wen, M., Di Sarno, L., Wen, C. and Hao, J. L., In support of circular economy to evaluate the effects of policies of construction and demolition waste management in three key cities in Yangtze River Delta. *Sustainable Chemistry and Pharmacy*, 2022. 26: p. 100625.

[24] Ma, M., Tam, V. W., Le, K. N. and Li, W., Challenges in current construction and demolition waste recycling: A China study. *Waste Management*, 2020. 118: p. 610-625.

[25] Rameezdeen, R., Chileshe, N., Hosseini, M. and Lehmann, S., A qualitative examination of major barriers in implementation of reverse logistics within the South Australian construction sector. *International Journal of Construction Management*, 2016. 16(3): p. 185-196.

[26] Ghaffar, S. H., Burman, M. and Braimah, N., Pathways to circular construction: An integrated management of construction and demolition waste for resource recovery. *Journal of Cleaner Production*, 2020. 244: p. 118710.

[27] Shooshtarian, S., Caldera, S., Maqsood, T., Ryley, T. and Khalfan, M., An investigation into challenges and opportunities in the Australian construction and

demolition waste management system. *Engineering, Construction and Architectural Management*, 2021.

[28] Shooshtarian, S., Hosseini, M. R., Kocaturk, T., Arnel, T. and Garofano, N., Circular economy in the Australian AEC industry: Investigation of barriers and enablers. *Building Research & Information*, 51(1): p. 56–68.

[29] Tam, V. W., Comparing the implementation of concrete recycling in the Australian and Japanese construction industries. *Journal of Cleaner Production*, 2009. 17(7): p. 688-702.

[30] Crocker, R. and Lehmann, S., *Motivating change: Sustainable design and behaviour in the built environment*. 2013, London, UK: Routledge.

[31] Udawatta, N., Zuo, J., Chiveralls, K. and Zillante, G., Attitudinal and behavioural approaches to improving waste management on construction projects in Australia: benefits and limitations. *International Journal of Construction Management*, 2015. 15(2): p. 137-147.

[32] Chapman, J., Skinner, N. and Searle, S., *Working towards sustainability: exploring the workplace as a site for pro-environmental behavioural change*, in *Motivating change: sustainable design and behaviour in the built environment*. 2013, Routledge: London, UK. p. 88-102.

[33] National Waste Policy, *Less Waste. More Resources*, W.a.t.E. Department of Agriculture, Editor. 2018: Canberra, Australia.

[34] Shooshtarian, S., Maqsood, T., Wong, P. S., Yang, R. J. and Khalfan, M., Review of waste strategy documents in Australia: Analysis of strategies for construction and demolition waste. *International Journal of Environmental Technology and Management*, 2020. 23(1): p. 1-21.

[35] Senate Environment and Communications References Committee, *Never waste a crisis: the waste and recycling industry in Australia*. 2018, Parliament of Australia Canberra, Australia

[36] Zaman, A., Caceres Ruiz, A. M., Shooshtarian, S., Ryley, T., Caldera, S. and Maqsood, T., Development of the circular economy design guidelines for the Australian built environment sector. *Sustainability*, 2023. 15(3): p. 2500.

Chapter 2

Implementation of the Circular Economy in Modular Construction

Salman Shooshtarian[1]
Peter SP Wong[1]
Tayyab Maqsood[1]
and Tim Ryley[2]
[1]School of Property, Construction and Project Management, RMIT University, Melbourne, Australia
[2]School of Engineering and Built Environment, Griffith University, Brisbane, Australia

Abstract

Modular construction is a building method where individual components, or modules, are prefabricated in a factory-controlled environment and then transported to the construction site. These modules are typically designed to fit together seamlessly to form a complete structure, such as a home or commercial building. The use of secondary materials in modular construction is a sustainable strategy to enhance products with recycled content (PwRC) uptake in the building and construction sector. The factory setting allows for greater exploration and application of eco-friendly materials, which may be challenging to implement in conventional construction environments. This chapter provides the findings of a literature review and a case study (ATCO Structures and Logistics) that provides insight into the use of PwRC in modular construction. The case study results showed that there are 16 categories of challenges that hinder the application of PwRC in modular construction. The interviews also revealed that ten strategies could further enhance the adoption of PwRC in modular construction. Lastly, according to the research findings, this chapter provides a framework

In: Policy, Practice and Education of the Circular Economy in Built Environments
Editors: Salman Shooshtarian, Peter SP Wong and Tayyab Maqsood
ISBN: 979-8-89530-623-9

developed to guide efforts in enhancing the uptake of PwRC in modular construction. This framework consists of four components that propose strategies to enhance organisational capacity, supply chain performance and recycled product quality as well as policies that can support this transformation across the first three components.

Keywords: financial benefits, products with recycled content, modular construction, public projects, challenges, client preferences and obligations

1. Introduction

Modular construction is a building method where individual components, or modules, are prefabricated in a factory-controlled environment and then transported to the construction site. These modules are typically designed to fit together seamlessly to form a complete structure, such as a home or commercial building. Modular construction comes in several forms. It can be delivered "as a box" in finished form, as a flat pack – separate parts such as wall frames and floor panels – or as a kit of smaller components. Prefabrication is revolutionising the construction sector at every level by bridging construction and manufacturing. This concept is encapsulated in the definition of industrialised construction, which involves the prefabrication, modularisation, and standardisation of construction processes and assets in controlled factory environments [2].

Modular construction has the potential to also transform project management for construction firms, with onsite construction projects completed in as little as 12 weeks instead of the traditional 12-month timeline. Additionally, modular design enhances their ability to handle multiple projects simultaneously without compromising quality or risking damage to their reputation. The use of modular construction provides several benefits as outlined in Table 1. These advantages encompass social, environmental, and economic aspects, offering a more sustainable construction approach overall.

In modular construction, bulk material orders allow manufacturers to fabricate multiple modules simultaneously, leading to lower prices from suppliers and reduced labour and transportation needs. This results in cost and time savings for the project. Additionally, prefabricated construction reduces the on-site labour force, lowering total labour costs by approximately 25% compared to traditional methods; this is specifically evidenced when the labour force has to be deployed in regional areas. It is well documented that when construction phase activities overlap with prefabricated construction, there is a reduction in construction time of approximately 40% compared to traditional construction methods [3]. One study focused on dwellings only, indicates that modular construction can cut construction times by as much as 50% and reduce costs by up to 20% [4]. Another analysis estimates claimed offsite construction is three times more productive than onsite [2].

Another key benefit is the immunity to weather disruptions, allowing for year-round production and the potential for multiple shifts, leading to optimal utilisation of plant and equipment. This high utilisation rate helps justify investments in new modular technologies. Furthermore, the controlled factory environment enhances safety, resulting in fewer days lost to injuries compared to onsite construction. Offsite construction also enables concurrent work processes, such as foundation and wall construction, which can then be integrated onsite. Additionally, the use of digital technologies is more extensive in a factory setting, providing greater precision in cost estimation and delivery times. Process control technologies and automation reduce time and waste, leading to fewer defects and the need for rework [5]. In terms of construction work safety, the adoption of this technology can provide a safer working environment which can result in the reduction of safety incidents by up to 80% [6].

This type of construction also offers environmental advantages, including decreased construction waste and CO_2 emissions, as well as reduced disturbance to neighbouring sites by minimising on-site noise and dust pollution. A recent report by the Waste & Resources Action Program (WRAP) shows that 90% of waste reduction can be obtained by optimising the use of off-site construction [7]. Roughly similar figures were found in a case study conducted in Australia [8]. The authors of this case study inferred that there is an inversely proportional relationship between the level of prefabrication and waste generation.

However, this construction technology comes with some limitations. As outlined in Table 1, there are seven primary limitations in modular construction, and these can multiply if modular construction organisations fail to adopt best management practices. Limitations exist at various stages of off-site construction, from planning and design to manufacturing and installation. This is inherent in all construction and not unique to modular construction however forward planning is strategically important as there is little time to figure out issues in this form of construction.

Table 1. Summary of the major advantages and disadvantages of modular construction

	Advantages	Description	Disadvantages	Description
Economic benefits	1. Faster construction	Modular construction involves simultaneous off-site fabrication and on-site foundation work, reducing overall construction time	1. Transportation costs	Shipping large modules to the site can be expensive
	2. Cost-effective (affordable)	Prefabricated modules are produced in a controlled environment which may result in reducing labour and material costs. Furthermore, with a wide range of designs and inclusions, the final cost of the project may be lower than expected, depending on the contextual factors.	2. Limited customisation	Modular designs often have limitations in terms of architectural creativity and flexibility due to the predetermined transportation height/width limitations
	3. Improved construction quality	Factory-controlled construction ensures higher quality control compared to traditional on-site construction. DfMA advantages are exploited for subassemblies reducing errors and quality hold points	3. Dependency on factory production and limitations on the material selections	Delays in the factory can lead to project delays
Socio-environmental benefits	1. Less disturbance to building site local residents' everyday lives	Less disturbance to the building site's neighbours by minimising on-site noise and dust pollution and project completion time	1. Size limitations	Restrictions on the dimensions of a prefabricated structure might exist due to regulations concerning the width of roads allowable for transportation.
	2. Reduced material waste	Precise manufacturing processes minimise waste or reuse/repurpose waste that would otherwise be disposed of	2. Potential for damage	Modules can be susceptible to damage during transportation and installation
	3. Safer working conditions	With much of the construction work done off-site, on-site workers are exposed to fewer hazards and other conditions such as weather, ground or crew politics	3. Requires skilled labour	Specialised skills & expertise are needed for module fabrication and on-site assembly and commissioning
	4. Post-disaster housing management	Rapid deployment of housing units to areas affected by disasters, providing quick and efficient temporary housing solutions for displaced individuals and communities	4. High running costs	It requires high investment costs as well as additional project planning

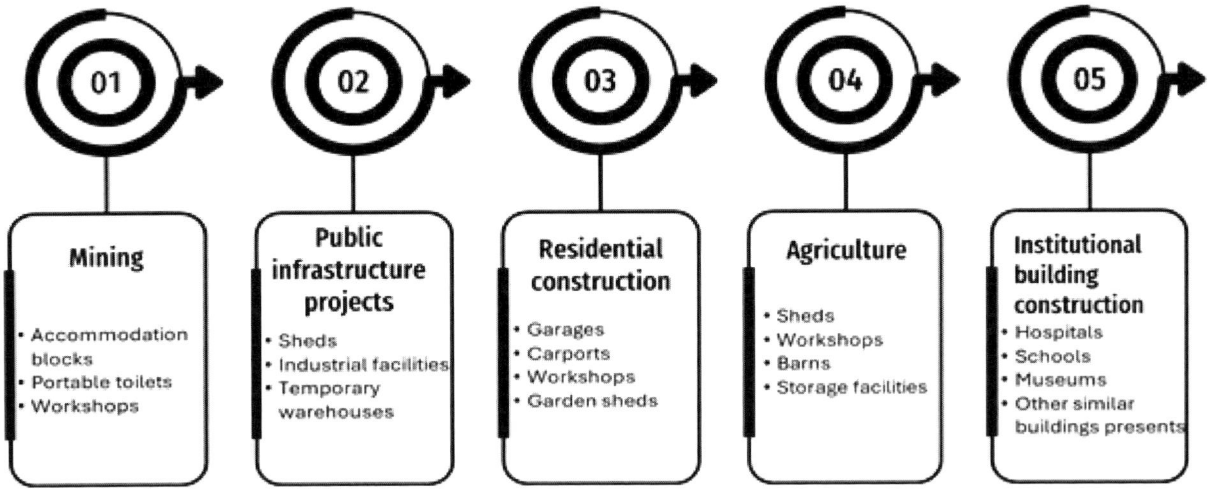

Source: Authors.

Figure 1. Major industries and sectors influencing modular construction and the relevant product examples.

The use of secondary materials in modular construction is a sustainable strategy to enhance PwRC uptake in the building and construction sector. The factory setting allows for greater exploration and application of eco-friendly materials, which may be challenging to implement in conventional construction environments. The following sections provide the chapter objectives, explore the research approach employed in this study and provide an overview of the application of PwRC in modular construction.

2. Objectives

This chapter addresses three objectives as shown below.

1. To explore the main benefits and motivations of integration of PwRC into modular construction products
2. To understand the main barriers and strategies to enhance the use of PwRC in modular construction
3. To develop a framework to guide the successful incorporation of PwRC in modular construction.

3. Research Approach

3.1. Review of Literature

This section of the chapter presents the research approach employed. First, the research team conducted an extensive literature review to offer insight into the status of modular construction uptake in Australia and explore the main drivers and challenges to enhance its uptake in the Australian context. This review also identified the benefits of using PwRC in modular construction in Australia. The sources used in this step included journal articles and government and industry reports that are freely available online. Care was taken to only capture the literature that describes the use of PwRC in recent years to reflect the contemporary efforts and advances in using these materials.

3.2. Case Study Approach

In the second step, a case study analysis was conducted to understand Australians' experience in using PwRC in modular construction. In consultation with the project partners and waste experts, the criteria for the selection of the case study were set. The case study had to meet the following criteria:

- Use of PwRC in modular construction;
- Its recent application in the organisation projects
- Ease of access to key stakeholders.

After identifying the project, ATCO Structures & Logistics, the perceptions of key stakeholder groups playing an essential role in using PwRC were sought through conducting a series of interviews. The interviewees were selected in consultation with the ATCO S&L sustainability team. An invitation email was sent to each stakeholder to arrange suitable interview times. The online interviews were conducted using the Microsoft Teams online platform. Each interview took a maximum of one hour and was recorded for transcription purposes. Three types of questions were asked during interviews: participants' details, their experience using the PwRC modular construction, and the enablers of and barriers to using such resources in this section of the industry.

3.3. Data Analysis

The data extracted from the literature review was thematically analysed preceding the creation of the major clusters of benefits, challenges and barriers to using PwRC in modular construction projects in Australia. The literature analysis provided the basis for a case study analysis in the Australian construction industry. Audio data captured from 6 interviewees were carefully transcribed by a professional transcriber using the word-for-word method prior to the quality verification of the text data by the research team. The transcripts were analysed using the NVivo Pro 12 application, which facilitates codifying text-based qualitative data. A thematic analysis method was applied to identify emerging themes that were related to the three research objectives. Furthermore, to better compare the qualitative data, quantitative descriptive analysis methods were adopted. The frequency of distribution of various categories of factors identified in the interviews was the main statistical measure used to conduct these comparisons.

4. Literature Review

Modular construction offers a wide range of environmental and economic benefits. This section of the chapter reviews these benefits in light of the use of PwRC and waste minimisation during their application.

4.1. Modular Construction in Australia: An Overview

All modern methods of construction (MMC) approaches particularly modular construction are experiencing a surge in popularity in Australia due to recognised value as an efficient construction method. As outlined before, this approach reduces overall construction costs, saves time, and encourages innovation.

With less than 5% of the total construction, the adoption of modular construction in Australia lags behind that of many other advanced economies [5], such as Scandinavian countries, Japan, Singapore and Germany. The PrefabAus industry roadmap report indicates that by 2025, this figure will reach 15%. While modular design may not be the perfect solution to address all the challenges in the Australian housing market, it does represent an initial step toward a fairer housing solution. One may consider the potential of high-quality buildings that can be disassembled, moved, and updated easily in a community that greatly needs more affordable housing options [9]. Furthermore, Australia boasts the largest average home size globally. However, modular housing, akin to the tiny homes movement, promotes the idea of living with a smaller footprint.

The public sector's capital expenditure is expected to increase by 2.8% in 2023-24, leading to a rise in demand for prefabricated building products such as sheds, industrial facilities, and temporary warehouses. This growth is driven by major infrastructure projects like the Western Sydney Infrastructure Plan, Perth's METRONET expansion, and the Victorian Government's Metro [10]. Federal and State governments across the nation are increasingly showing significant interest in modular construction. For instance, the Department of

Foreign Affairs and Trade (DFAT) requires the use of modular building solutions for replacing assets in the Pacific Islands, as well as for critical consular and diplomatic facilities. Acknowledging the security benefits of offsite construction, the Australian modular construction industry has supplied the majority of construction needs for the Australian Defence Forces over the past decade [5].

At the state level, Queensland is committed to extending its apprentice-led QBuild program, producing prefabricated homes in a Brisbane factory. Australian prefabrication supplies 80% of the build for the 2032 Queensland Olympic Games athletes' village. Correctional Services New South Wales (NSW) has pledged to boost the production of modular homes by June 2025, aiming to enhance the supply of Indigenous housing. Meanwhile, in Victoria, the state government has started providing modular housing to individuals at risk of homelessness [5]. The major external industries and sectors that influence the performance of the modular construction industry in Australia include mining, public infrastructure construction, private residential construction, agriculture, and institutional building construction (Figure 1).

A recent report by IBISWorld [10] suggests a positive growth in demand from institutional building construction, public infrastructure construction, and mining sectors in 2023-24. However, a decline is expected in private residential construction and agriculture sectors.

PrefabAUS [5] estimates indicate that the Australian modular construction industry has the potential to yield annual benefits of $9 billion. This estimation is based on the assumption of a 20% cost saving and the nation achieving a 30% share of modular construction in the construction market by 2033. According to a recent report by IBISWorld [10], the prefabricated metal building manufacturing industry is projected to experience an annual revenue growth rate of 1.7% over the five years through 2028-29, reaching $3 billion.

When it comes to support for the Australian modular construction industry, two key factors stand out. The industry benefits from free trade agreements with China, Vietnam, and New Zealand. These agreements exempt prefabricated metal building products from the general tariff of 5.0%, making exports to these markets easier. Over the past five years, these trade deals have helped industry exporters expand their access to international customers. Additionally, manufacturers are protected by anti-dumping laws, ensuring fair competition and safeguarding against unfair trading practices. PrefabAUS is an industry association dedicated to promoting and supporting the use of prefabricated construction methods to enhance construction efficiency, sustainability, and affordability. The organization serves as a bridge between manufacturers and government agencies, advocating for the sector's interests and needs. Through collaboration with relevant bodies, PrefabAUS plays a crucial role in influencing policies, regulations, and standards affecting the modular building sector.

4.2. Drivers of Adopting Modular Construction in the Residential Sector

The application of this construction methodology in Australia is driven by five major factors (Figure 2). Modular housing in Australia has been advocated as both an immediate response to accommodate those who are affected by natural disasters (short term), and a rapid method to deliver permanent residences, aiming to alleviate the nation's housing shortage (long-term) and a solution to the limited supply of construction labour (medium-term) [5]. Recently, modular housing has been used as an efficient solution for flood relief in the Northern Rivers of NSW and central west towns such as Eugowra.

In Australia, limited construction skilled labour presents an opportunity and incentive to quickly scale up the prefabrication industry for success in the residential sector. Furthermore, Australia is known for paying some of the highest construction labour wages globally. Factors contributing to this include Australia's high cost of living, strong labour unions, stringent workplace health and safety regulations, and the nation's resource-rich economy, which drives up wages in industries like construction.

Disaster relief solution	Short term response to post disaster management
Solution to housing crises	Permanent family homes to address housing shortage
Limited construction labour	Compensating for ongoing skilled labour shortages with increased productivity rates
High construction wages	Streamlining labour processes and reducing the need for extensive on-site work
Increased interest and investment	Recent increased interest by public and private sectors in modular construction application and manufacturing

Source: Authors.

Figure 2. The main drivers for using modular construction in Australia.

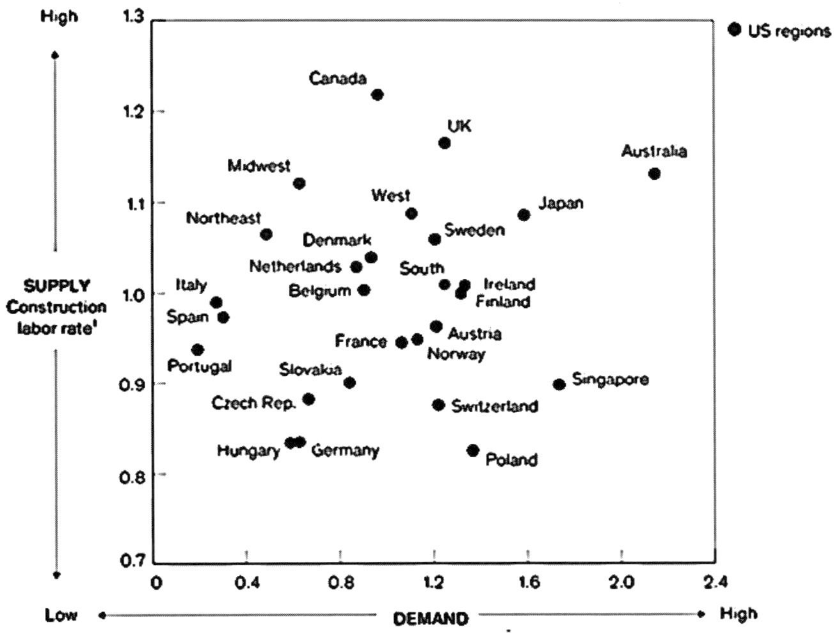

Source: Bertram, Fuchs [4].

Figure 3. Near-term demand for new housing against construction labour supply.

The majority of Australian builders are small businesses with one or two people, relying heavily on subcontracted and agency labour for various stages of building projects. A recent report published by IBISWorld [10] indicated that the COVID-19 pandemic severely disrupted supply chains for building materials, fuel, capital equipment and skilled labour; delaying building progress and escalating input prices and wage rates in Australia. The use of modular construction can compensate for the skilled labour shortage as it provides higher rates of construction productivity. Since unmet housing demand and the relative scarcity and cost of construction labour are the major predictors of where modular construction can gain traction, a report by McKinsey & Company [4] showed where these two factors intersect as depicted in Figure 3. This report highlights the potential growth in markets such as Australia, the UK, Singapore, and the US regional.

There has been a notable rise in interest and investment in modular construction recently. This increase is largely attributed to the Australian government's commitment to procuring prefabricated construction structures.

4.3. Barriers to Adopting Modular Construction in the Residential Sector

Barriers to adopting this methodology in Australia come in different forms and at various levels of impact. There is limited knowledge of modular construction applicability nationwide. The industry model in Australia is primarily focused on basic and remote housing, with prefabrication largely limited to the basic definition of offsite production. However, the industry has yet to fully embrace new processes and models on a larger scale. In many cases, Australian prefabrication simply transfers traditional construction methods to a factory setting [5]. This is of particular importance, as the scale is needed to justify investment across off-site construction manufacturing and application. Figure 4 provides a summary of the major barriers to using modular construction in Australia.

Establishing modular manufacturing incurs significant capital costs. The initial investment is required to set up the infrastructure, equipment, and processes necessary for efficient and effective production of modular components. As outlined before, the pandemic-related disruptions have added to the complexity and contributed to higher setup costs. This together with the uncertainties about the return on investment (ROI) in the industry has made setting up modular construction facilities more challenging and costly, posing a major hurdle for the industry in Australia.

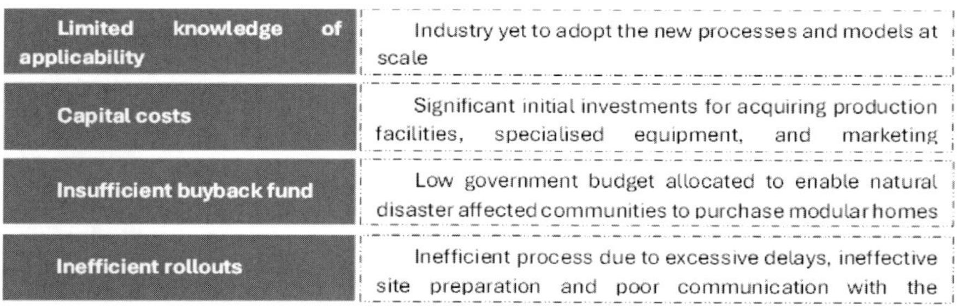

Source: Authors.

Figure 4. The main barriers to using modular construction in Australia.

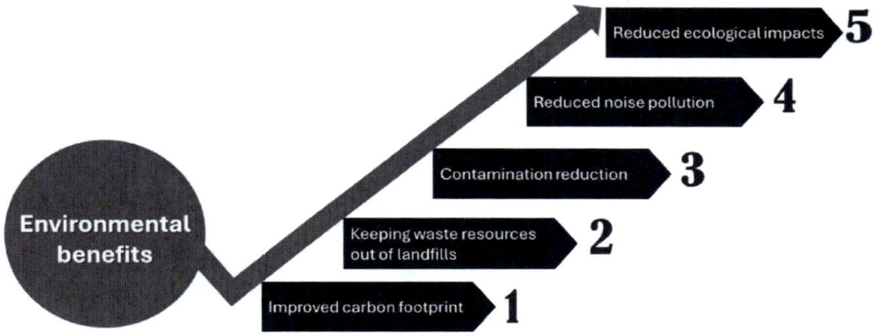

Source: Authors.

Figure 5. The major environmental benefits of using PwRC in modular construction.

The complexity of using modular construction as a disaster relief solution is intertwined with government initiatives. In a recent article published in the Guardian [11], three major barriers to the application of modular construction as a disaster relief solution were reported to be the lack of community consultation, small funding pools for buyback initiatives, and inefficient rollouts. According to local news [12], the buyback initiatives in Australia may not offer property buybacks at pre-disaster values for all disaster-affected communities. Having access to this fund would have been beneficial for these communities to raise their homes, especially those considering modular construction as a quick accommodation solution. The issue with inefficient rollouts in natural disaster-affected areas is linked to the delays in providing modular temporary housing solutions, ineffective construction site preparation and poor communications with the stakeholders [11].

4.4. Environmental Benefits of Using PwRC in Modular Construction

As visualised in Figure 5, there are five major environmental benefits with respect to using PwRC in modular construction. The first environmental benefit of this approach is carbon emission savings. Carbon savings primarily result from the reduced consumption of natural resources in the building and construction sector due to the utilisation of PwRC. Carbon emission saving can be estimated using a range of available metrics including global warming potential (measured in t CO_2 eq) [13]. The findings of an empirical research study [14] indicate that modular construction reduces embodied energy by

56% and 26% in small and large structures, respectively. Technically, this can offset the additional carbon emissions arising from waste recycling, thus improving the environmental performance of PwRC.

The use of secondary materials in modular construction occurs in two different forms: recycled materials and nonrecycled materials. A research study in Australia [13] showed that reusing waste from non-recycled resources has a superior carbon footprint than recycled materials. This is because some PwRC are invariably wasted or contaminated in the process, and the recycling process requires energy consumption. In some cases, construction and demolition waste from a site can be reused in the new construction, reducing the necessity to transport these materials to and from recycling facilities. A study conducted in Australia [15] indicated that reusing demolition waste in the new construction on the same site has resulted in 1,000 fewer truck movements, 9,000 L of diesel saved, and therefore, 663 t of CO_2 were avoided. The study found that the carbon emission reductions were achieved in three key areas: onsite, through the replacement of raw materials with recovered demolition products; offsite, through recycling; and during transport, through reduced truck movements.

The second benefit is related to reduced waste disposal from demolition activities. Modular designs enable not only offsite construction but offsite demolition as well. A University of Melbourne's Centre for Advanced Manufacturing of Prefabricated Housing report [16] suggests that all components of modular construction can be reused, thus eliminating the waste that would typically end up in landfills.

The third advantage focuses on minimizing the risk of waste resource contamination at the end of their service life. Modular constructions are usually designed for easy disassembly, reducing the contamination of separated waste resources and extending their lifespan for second or subsequent lifecycles. Additionally, modular construction processes reduce the impact of potentially environmentally hazardous waste materials, which are more challenging to manage compared to non-hazardous construction and demolition waste materials. The fourth advantage of using PwRC in modular construction is the reduction of noise pollution. In the environmentally controlled setting of an off-site factory, the process of repurposing and applying resultant PwRC in prefabricated building elements can be completed with minimal environmental noise. This is in contrast to on-site construction, where the use of materials and construction methods can lead to significant noise disturbances for nearby residents and businesses. By conducting these processes in a controlled factory environment, noise pollution can be greatly reduced, contributing to the sustainable application of these PwRC. The last environmental advantage of this application is the reduced impacts on the ecology. Since the majority of application of PwRC in a modular approach takes place off-site, it significantly diminishes the ecological impact including reduced soil erosion and less harm to the local ecosystem.

4.5. Economic Benefits of Using PwRC in Modular Construction

The use of PwRC in modular construction can provide multiple economic benefits (Figure 6). The first benefit is the decreased requirement for rework caused by building defects. Off-site construction allows for the highest quality of construction, leading to enhanced implementation of PwRC with minimal waste from defects during both construction and operation. This results in cost savings for constructors and building owners by avoiding the costs associated with rework. Moreover, when constructed to withstand earthquakes and high wind loads, they often outperform conventional structures designed to resist these forces [3]. This advantage can reduce externally induced defects in prefabricated components that incorporate PwRC.

The second benefit involves extending the PwRC lifecycle. Although an increasing portion of modular construction focuses on permanent structures (similar to conventional buildings), many businesses continue to utilise modular buildings for temporary purposes. Relocatable buildings are brought to sites for rapid expansion or as temporary facilities for students, patients, or employees during renovations or emergencies. Once the additional space is no longer required, the modular building is taken off-site and repurposed for future use. This indicates that PwRC can continue to function as a construction material, reducing the need for demolition and greatly enhancing its circularity.

The third benefit is in connection with a more standardised (industrialised) approach to using PwRC reduces the complexity of utilising these waste-based resources. This streamlined approach enhances their application, leading to significant cost savings. Furthermore, as the prefabricated modules are repeatedly produced for clients, supplies can be ordered and stored onsite. This enables any PwRC leftover resources from one project to be used in subsequent modules, enhancing their utilization efficiency, especially during supply shortages. Moreover, with the increasing interest in prefabricated buildings, the use of PwRC in this industry can tap into a new and sustainable market. This, in turn, fosters an economy of scale that ultimately enhances the economic competitiveness of PwRC.

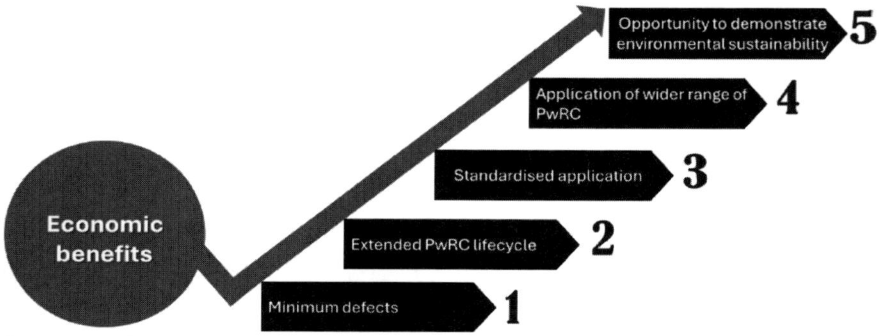

Source: Authors.

Figure 6. The major economic benefits of using PwRC in modular construction.

The fourth benefit focuses on the application of a wider range of PwRC in an industrialised environment. Constructors can explore innovative ways to incorporate various PwRC that may not be feasible in onsite construction due to constraints such as time and space limitations or safety requirements. The increased demand for a wider range of PwRC can attract investment in advanced repurposing technologies and the development of new supply chains. This, in turn, can lead to a reduction in the production cost of these resources.

The last benefit involves capacity building within the modular construction industry and organisations to enhance their environmental sustainability and showcase it effectively. In many countries, including Australia, environmental sustainability initiatives such as Zero Waste movements urge the government and the industry to take a more circular approach. This is now clearly reflected in the decision-making processes of organisations involved in the building and construction sector. In such scenarios, these organisations are incentivised to utilise PwRC to showcase their sustainable practices in business operations [17] and to be recognised by various environmental sustainability rating schemes [18]. This will increase the likelihood of securing additional construction contracts that prioritise environmental sustainability. In a recent study conducted in Australia [15], the research participants indicated that the need for organisations to ensure competitive advantage and future-proofing was identified as the second main motivation for using PwRC in construction projects.

Table 2. ATCO S&L's divisions of operation

Division	Description
Modular buildings for hire & sale	Accommodation buildings, daycares and schools, kitchen and diners, commercial offices, parks cabins, classrooms, recreation and gymnasiums, site offices and demountable, toilets and ablutions, and trailer mounted units
Atco villages	Furnished accommodation for workers in the construction, mining, education, oil and gas and other commercial industries
Manufacturing capability	A wide range of modular products such as blast-resistant modular buildings, customised modular buildings, education facilities, residential, commercial, switch room and control rooms, office accommodation, and accommodation villages and camps
Project construction services -	design and engineering, project life cycle and control, master planning, civil works, project management, contracting, pre-engineered buildings, sport and recreational facilities, water and wastewater management, power generation, building approvals and certification, service and maintenance, contracting, and logistics management (road and shipping)

5. Case Study Findings

5.1. Background to the Case Organisation

ATCO Structures and Logistics (ATCO S&L) is a private company that earns most of its revenue from the commercial and industrial building construction sector. It is engaged in the construction, sale, and hire of transportable buildings. In 2024, this manufacturing organisation is projected to hold 6.5% of the market share for prefabricated metal building manufacturing, ranking it as the second-largest manufacturer in Australia in this category [10]. The company operates sales offices and manufacturing facilities throughout Australia with a particular focus on Queensland, Victoria and Western Australia through the divisions outlined in Table 2.

The main ATCO S&L modular products include commercial offices, government facilities (e.g., emergency, health and custodial services), sporting facilities and educational facilities [1]. These products are (custom) designed, manufactured, transported and installed by ATCO S&L either temporarily or permanently (Figure 7).

Source: ATCO Structures and Logistics [1].

Figure 7. Top left (office building), top right (educational building), bottom left (sporting facility), bottom right (correctional facility).

5.2. Organisational Environmental Sustainability

ATCO S&L data shows that PwRCs are utilised in diverse building elements, including base steel, base, roof, wall, and window materials, as well as lifters. The proportion of recycled content varies, with roof, wall, and window materials containing 17.4%, base materials 25%, and lifters and base steel containing as high as 98% recycled content.

6. Application of PWRC in ATCO S&L

6.1. Descriptive Findings: Participants Profile

Six key stakeholders who influence the use of PwRC in modular construction products were interviewed between March and March 2024 (Table 3). The interviewees were experienced individuals with a significant track record of involvement in modular construction in Australia. Five of these interviewees were employed in different departments of ATCO S&L and were based in Queensland at the time of the interview.

Table 3. Summary of participants' profiles in the case study

Stakeholder type	Experience/expertise	Participant code
Technical Manager	25 years of experience in various sub-fields of MMC in Europe, North America, Africa and Australia. Management of engineering and compliance aspects of the organisation's modular products.	C_1P_1
Design Manager	More than 15 years of experience in the Australian MMC industry with responsibilities including detailing, designing and estimation	C_1P_2
Operations Manager	More than 25 years of working experience in the Australian MMC industry.	C_1P_3
Branch Manager	Extensive experience working in the study organisation's business development sales departments.	C_1P_4
National HSEQ* Manager	A manager with experience in different fields and employment sectors including work health and safety, construction markets and education and training.	C_1P_5
Supplier	Representative of supplying organisation (steel) with expertise in sustainability and extensive experience in the steel manufacturing industry.	C_1P_6

*HSEQ: Health, Safety, Environment and Quality.

The technical manager of the study organisation (C_1P_1) was a design engineer by background with extensive engagement in a wide range of MMC activities on a global scale. He also managed the compliance and engineering aspects of the organisation's modular products. The design manager of ATCO S&L (C_1P_1) had worked for the organisation for more than 10 years, with the overall responsibility of designing modular buildings. The operations manager (C_1P_3) had also extensive experience in the building and construction sector particularly in the MMC industry. The organisation branch manager (C_1P_4) had stayed with the ATCO S&L for 14 years, with a speciality in business development and sales and marketing. The Health, Safety, Environment and Quality (HSEQ) manager of the organisation (C_1P_5) had expertise in various fields and employment sectors. The last interviewee (C_1P_6) was the representative of an organisation that supplied recycled steel to ATCO S&L.

When asked about their history of involvement in C&D waste management in their current or previous roles, the participants reported varying levels of exposure to this field. C_1P_1 primary experience in this field relates to the design engineering of modular components to reduce waste generation. C_1P_4 indicated that ATCO S&L is involved in waste minimisation and reusing the waste, to limit the waste of their products. C_1P_5 reported that his exposure to the waste management industry was through various contracts he had with organisations both in the industry and outside of that. He also mentioned that he is involved in managing waste in ATCO S&L. C_1P_6

mentioned that their organisation is involved in the recycling of metals in Australia. The rest of the participants did not have any direct involvement with waste management, as the modular construction industry does not typically generate too much C&D waste during manufacturing, transport, installation or operation.

6.2. C&D Waste Management and Procurement of PwRC

This section outlines the organisation's involvement in C&D waste management and the integration of PwRC in modular building products. It begins by detailing the organisation's strategies for managing C&D waste resources, followed by an overview of their initiatives for reusing modular products and procuring PwRC for new construction and retrofitting projects. Additionally, the contributions of each participant involved in these activities are highlighted where applicable. While some participants were directly involved in the procurement process, others were not, resulting in varying levels of understanding of the organization's procurement procedures among different participants.

Typically, the off-cuts and waste materials generated during the manufacturing and installation of modular buildings are disposed of in recycling bins. These waste resources are managed through three potential pathways: 1. Collection by third-party waste operators, 2. Reuse in the production of new building products or applications in the retrofitting projects, or 3. Disposal in landfills. To minimise waste generation, the organisation set a criterion for material procurement focusing on the 'packaging materials' issues. C_1P_1 reported that in some cases, the organisation negotiates with the supplier to reduce packaging waste materials. However, in most cases, the off-cuts and other waste materials are dumped into recycling bins, where, as outlined above, they might be reused in ACTO S&L products. An example of this reuse is for framed decks,

> 'Framed decks are typically lined with a structural ply lining on top...we may find that that structural ply is kind of weathered. So we will take that part off the deck…But we will reuse the deck frame, and reline it so that we don't have to continually go and get more steel framing built for our jobs. And we can continually reuse that' [C_1P_4].

Reuse *Materials that are reused*

- **Steel in framed decks**
- **Roof purlins**
- **Aluminium treads on stairs**
- **Handrails**

Repurposed *Materials that are reporuposed to be used in construction*

- **Solid core concrete:** Repurposed to be used in small projects
- **Steel pieces generated from punching frames:** Repurposed to packers to be used for tabbing and welding in the factory
- **Plywood offcuts:** Repurposed to packers to be used for the setting up of windows.

Supplied *Materials with recycled content that are supplied to ATCO*

- **Angle channels**
- **Flats and hot-rolled sections**
- **Steel that is used in modular components e.g. cladding**
- **Vinyl products**

Source: Authors.

Figure 8. Examples of reusing construction materials in the Case Study organisation.

Other materials that are reused in this organisation are presented in Figure 8.

In this section, the organisational procurement model is reviewed and some of the procurement criteria being used are presented. Overall, despite having slightly different procurement criteria, the PwRC procurement model is not different from other alternatives (C_1P_5), and their utilisation is to a limited extent as indicated by C_1P_3,

> 'The process of procuring waste materials, it would be no different to procuring any material' [C_1P_5].

> 'We don't tend to use a lot of recycled products if I'm being honest. When we do it, it would usually be in line with the tender requirements…occasionally we use recycled decking products and decking boards. That's about it.' [C_1P_3].

According to the interviews, there are several factors influencing the procurement of PwRC in ATCO S&L products. These factors are listed in Figures 11 and 12. It also appears that there are currently no specific criteria established for the procurement of PwRC. However, the most commonly cited criteria for PwRC procurement include cost, availability, associated risk, and suitability. C_1P_2 reported that

'It's still more like in a pipeline to look at those recycled materials, but again it all comes down to what's the cost involved in those materials, the availability of those materials and if there's any risk' [C_1P_2].

In ATCO S&L, generally, the design team specifies the construction materials based on the individual designs, which will be passed on to the manufacturing sector to procure. For the procurement of PwRC, these are suppliers that reach out to the organisation to sell their products. They share their products' specifications with the organisation (i.e., compliance team) to review them for fit for purposes point of view of the organisation.

'There are some products that now we've been introducing, that we're investigating to use as we can see a lot a few of, a couple of sales reps from different recycling companies, they're start coming into the business and introduce the product' [C_1P_2].

Most products are supplied through shorter supply chains with whom ATCO S&L established business arrangements. For instance, C_1P_6, the supplying organisation representative, indicated that they supply steel construction products/materials, some with a high level of recycled content to ATCO S&L. These materials include smaller angles channels, flats and other hot rolled sections (Figure 8). The participant also commented on the upstream material supply chain. They primarily purchase metal waste materials from demolition companies who in turn extract them from building demolition projects. Additionally, they source them from the manufacturing businesses, their manufacturing unit, and the general public.

'So anyone can come in with a trailer or a boot full of scrap metal, you know, not just steel, but copper, aluminium and other metals, and they would be purchased over the counter' [C_1P_6].

There are four reasons why the organisation tends to choose a shorter supply chain when procuring materials, these include (1) the trust established between parties to avoid falling into the greenwashing trap, (2) avoiding product variability (3) more control over meeting product quality requirements, and (4) lower prices.

'It was always to shorten the supply chain because you can reduce obviously variability, but you can also get closer to what you want rather than what you're given…and then usually that ends up reflecting in a lower price' [C_1P_1].

In some cases, the procurement is carried out by other suppliers for specific products e.g., timber products. For example, C_1P_1 indicated that timber products are sometimes sourced from outside of the shorter supply chain arrangement:

'Sometimes substitution comes by via an established supplier, and sometimes they come from different sources and so, for example, we have a supplier who supplies timber and timber products and they source them from all over the world' [C_1P_1].

6.3. Sustainability Benefits

In this section, the sustainability advantages of incorporating PwRC in modular construction, as perceived by the research participants, are thoroughly examined. The insights from the interviews provide a valuable complement to the literature review findings discussed. As illustrated in Figure 9, the participants identified ten key benefits of such an application that enhance social, environmental, and economic sustainability outcomes in modular construction.

6.3.1. Less Damage to Natural Resources & Pollution

In participants' views, the use of PwRC in modular construction can enhance the environmental sustainability of the building and construction sector. The reduced need for extracting virgin materials contributes to environmental protection and pollution reduction.

'Recycled products are better than, say, chopping down trees and creating decking boards and Poly ply and whatever else.' [C_1P_3].

'It's reducing even the amount of raw product we're taking from the planet, raw deforestation for mines' [C_1P_5].

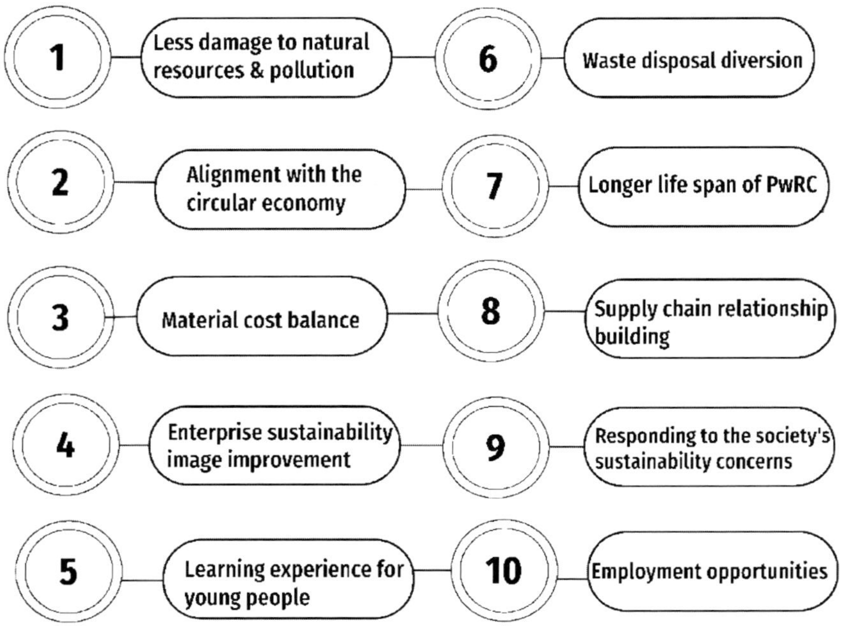

Source: Authors.

Figure 9. The major benefits of using PwRC in modular construction.

6.3.2. Waste Disposal Diversion

The application of PwRC can generate demand for waste resources, which in turn reduces the volume of waste sent to landfills. Waste disposal diversion results in improved economic and environmental sustainability in the sector.

> 'Where we're using recycled products is a great thing because it's reducing landfill, it's reducing microplastics in the ocean' [C_1P_5].

> 'Diversion of waste away from a waste stream back into a recycling stream' [C_1P_6].

6.3.3. Alignment with the Circular Economy Concept

Within the circular economy concept, the use of PwRC is emphasised as a key strategy to maintain the value of materials in the economy for as long as possible. According to C_1P_1, utilising locally sourced PwRC can further enhance the sector's circularity by reducing the need for transportation and energy consumption.

> 'If the market makes that circle of the circular economy as small as possible, rather than sending, your waste all the way to either China or India or somewhere in Europe or North America to then, be recycled there and probably reused there.' [C_1P_1].

> 'That's [using recycled products] the idea of how you can do more with less. The idea of generating less waste is important, but use less to, use less energy, generate fewer emissions, increased circularity, increasing material efficiency' [C_1P_6].

6.3.4. Longer Life Span of PwRC

One participant indicated that some PwRCs have better durability compared to products created from virgin materials. According to C_1P_4, This in the long run can provide an economic advantage for the manufacturer of modular products.

> 'A lot of these recycled products have a longer life span than the other ones. So, whilst they may be a higher initial investment, they certainly provide better economical advantages over the long term' [C_1P_4].

6.3.5. Material Cost Balance

As the consumption of natural resources continues to rise across various industries and for a range of products, their prices will likely increase. Additionally, in certain applications, the use of PwRC may not be technically feasible, leading to intense competition for virgin materials. Therefore, in the future, incorporating PwRC in modular construction could play a significant role in enhancing the economic sustainability of the building and construction sector.

> 'If we're not using recycled products, the cost of a raw product style material will start to outweigh the cost of recycled material because there's less and less of it' [C_1P_5].

> 'All of a sudden you're competing for those raw minerals because they're going into producing solar farms and wind farms which means there's less for general steel production, etc.' [C_1P_5].

6.3.6. Supply Chain Relationship Building

Given the reasons mentioned above, the use of sustainable materials is expected to increase across various industries. Therefore, it is prudent to begin adopting PwRC, which will, in turn, help establish an effective supply chain for these products. Working with an established supply chain will enhance the economic sustainability of the modular manufacturing industry. C_1P_5 indicated that the earlier development of such supply chains is a key to success in this space.

> 'So, I think from an economic point of view, the earlier you move into that, the earlier you secure supply chain, the earlier you build those relationships, it's going to be better for you when that cost economy starts to change' [C_1P_5].

6.3.7. Enterprise Sustainability Image Improvement

The use of PwRC can enhance the enterprise's social image of businesses operating within the building and construction sector. In the current climate, where the sector faces severe criticism for unsustainable practices, the application of these products not only supports social sustainability but, according to C_1P_4, can also offer economic advantages by partnering with organisations that either fund sustainability-focused projects or seek to market themselves as sustainable businesses.

> 'I guess that it comes down to ATCO's place in the world and ATCO being seen to be using recycled materials and being seen to be green and reducing your carbon footprint' [C_1P_3].

> 'but one of our competitors was openly promoting themselves and demonstrating that they were using recycled steel ... And then you've got a big, $500 million project that's being funded by the federal government, managed by a tier one company, which tier one company is going to go with ATCO over the competitor that's marketing themselves and has that positive social influence?' [C_1P_5].

6.3.8. Responding to the Society's Sustainability Concerns

Societies worldwide, especially in developed nations, have gained greater awareness of environmental sustainability. As a result, there has been a significant increase in demand for sustainable practices and policies in different industries and sectors over the recent years. The use of PwRC can contribute to the overall sustainability of the sector, effectively addressing this growing demand.

> 'We can see socially that society is pushing more and more and more for sustainability. They want green energy; they want less products and pollution. They want the electrification of the grid. They want to know that the people that they're getting things from are having the least amount of impact on the Earth for their future generations and for their own health. So, I think from a primary social area, it really comes down to how you market yourself and remain at the forefront of people's minds that are potentially buying your products' [C_1P_5].

6.3.9. A learning Experience for Young People

According to two participants (C_1P_1 and C_1P_2), using PwRC in modular products, especially those used in educational settings, can offer valuable learning opportunities for young people. This exposure allows them to understand the application of these products in buildings and similar modular structures, fostering a mindset aligned with sustainability principles. Consequently, this can further enhance the social sustainability of the sector.

> 'Daughter's school, they were having modular buildings installed and like everything else on a school site, gets turned into a learning experience for the children...and the story, where you're able to tell the story of a, particularly a modular constructed building, probably as a narrative' [C_1P_1].

> 'The education sector across Australia is the driving force behind introducing new recycled materials. It's all about that, helping the environment and things like that' [C_1P_2].

6.3.10. Employment Opportunities

The process of converting waste materials into products often involves several steps that are labour-intensive, creating employment opportunities compared to waste landfilling. Therefore, it can be concluded that utilising PwRC in modular construction offers significant benefits for social sustainability through triggering waste recovery activities that employment associated with them.

'It provides an economic return to those who return it to us as recyclers, the demolition companies, the manufacturers or, you know, …there's a social impact around that from an employment point of view' [C_1P_6].

6.4. Challenges of Using PwRC in Modular Construction

The number of challenges identified as hindering the application of PwRC in the modular construction sector far exceeded our expectations. After consolidating some challenge categories, 16 categories emerged. These categories include a broad range of issues, including supply chain, cost, organisational limitations and technical complexities. Figure 10 illustrates these ten categories. The most mentioned challenge category among these is 'cost complexities' followed by the 'industry conservative approach' and 'availability of PwRC.'

The following sections provide insights into the industry perceptions regarding these challenge categories; they are presented in no specific order. The challenges highlighted in this section offer a foundation for advancing the modular construction industry, particularly in enhancing the incorporation of PwRC in modular products.

Figure 10. The main challenges of using PwRC in modular construction.

6.4.1. Project Characteristics

The quality and quantity of PwRC utilisation in construction projects depend on various factors related to the project's nature, including both physical and non-physical aspects. These factors can hinder or complicate the use of PwRC in construction projects. As outlined by C1P4, these factors can include the building's age, the project's size, and the allocated budget.

> 'So, we have buildings that are two months old, and we have buildings that are 20 years old. And obviously, the specification between those two buildings varies a lot' [C_1P_4].

> 'But we can't really, on bigger jobs, it's harder for us to use those [PwRC] because our clients will go, "Well, hold on a second, this is a three-and-a-half million-dollar job, why have I got second-hand materials here? What's this doing here? What's that doing here?' [C_1P_4].

> 'Whether they've got a tight budget or a loose budget, whatever that may be...whether their project can allow for it that will determine whether they can take it or not' [C_1P_4].

6.4.2. Cost Complexities

The issue of cost complexities was cited by several research participants. This issue has three dimensions and can be divided into three subcategories based on the interviewees' responses: 1) The higher cost of PwRC compared to conventional materials, 2) Market interventions by competitors against PwRC supply, and 3) The client's willingness to pay extra for PwRC.

> 'In order to be marketable it would then have to have all of that testing, and when you amortise all of that testing into a product, becomes more expensive, particularly when you're trying to introduce it into a new market' [C_1P_1].

> 'The exciting and the funny side of this is that you know, everyone's supposed to be recycled should be cheaper. But it's vice versa, right? When it comes to recycling the cost of the materials it's almost double the cost of the original, it's not that cheap [C_1P_2].

> 'Generally, it's, it tends to be those recycled products are generally more expensive than standard, say timber products, particularly with decking boards' [C_1P_3].

'If sustainability is number 1, then yeah, we're looking great. However, if the price is number 1, then those options can look quite expensive in the short term' [C_1P_4].

'There is an element of cost competitiveness and I think that relates a lot more to new products in the market versus established players, particularly when there might possibly be a premium being paid because of the lack of competition, which when competition comes in business...I'll say that business readjusts itself to suit the market, to maybe make it harder for a new entry to come into that market as well' [C_1P_1].

'So there's the initial upfront capital cost interrogation of product for product comparison. Then there's the life of asset cost incorporating maintenance and replacement over a 15 or 20-year product life cycle' [C_1P_5].

'Five, six, seven years ago, there used to be a push for air conditioners to have a certain star rating, for example, for their sustainability. Now, we would have to charge them to put new air conditioners into buildings so that they have that star rating and as soon as they saw the price of that, the answer was always, well, no, don't worry about it' [C_1P_4].

6.4.3. Industry Conservative Approach

The building and construction sector is often viewed as resistant to innovation and change, with a strong aversion to the risks that come with any change. The industry's supply chain processes have evolved over decades, and any request for change is seen as disruptive, leading to various complications. This resistance is especially evident among senior managers, who tend to rely on familiar supply chains and products. C_1P_1 described conservatism as a sign of ignorance. Below are some comments from participants on this issue:

'The construction industry, by its nature, is very conservative. But, like I said, a lot of things are based on, "It worked last time. I will do it again" ... But I think it's a lack of imagination, is probably another way of looking at it and using a conservative, conservatism in the way of ignorance' [C_1P_1].

> 'But when it comes to the modular industry, I would say 90% of the buildings, that use in the modular industry are they all using James Hardie products, for instance, or Colour, or Colourbond metal cladding' [C_1P_2].

> 'Senior business managers in the larger modular firms, have spent most of their career in those businesses, so they're so comfortable and familiar with what they've done and their supply chain… The habitual mentality of, "We've always used this product. We're comfortable with this product. We know this product. We know this supplier'[C_1P_5].

6.4.4. Greenwashing and Product Misrepresentation

This category refers to a well-established concept in the recycling practice and research. It is the practice of misleading end users by falsely claiming or exaggerating the environmental benefits or recycled content of a product. One participant (C_1P_1) suggested that greenwashed products could pose a challenge to the seamless acceptance of PwRC in the sector.

> 'There are some that have been green-washed and placed into the market as well…[they] are sold as green that essentially aren't, and they're still charging a premium or the selling a product that is not necessarily fit for purpose in terms of its longevity or application in our particular builds' [C_1P_1].

> 'They all currently have drawbacks in either their credentials as being completely recycled material. Having recently done some testing on some materials where they say, "Oh, it was made from this and this," and so due diligence and only a few hundred dollars later, I can get a test done and have it put to them that it's actually not what you say it is. So you actually misrepresenting your product' [C_1P_1].

6.4.5. Limited Availability of PwRC and an Immature Supply Chain

The limited availability of PwRC was a concern raised by three participants, who highlighted its significance given the volume and industrialised nature of modular construction. They pointed out that in modular construction the components must be readily available and supplied to manufacturing sites on time to prevent any disruption to the sequential production processes. One participant (C_1P_1) highlighted that the availability of some PwRC is constrained by the comparatively small and immature supply chain operating in Australia, which serves a market of 25 million people. Below are some responses from the participants:

> 'An Australian supply chain that is only supplying a population of 25, 26 million people. So there are certain products that I know that are available elsewhere that are not available in Australia and there's barriers to entry for that as well ... Some of the fire-rated products, some of the fire-rated board materials and stuff like that, readily available in the UK and Europe and [...] in North America' [C_1P_1].

> 'But again it's [PwRC application] all come down to what's the cost involved into those materials and the availability of those materials and if there's any risk' [C_1P_2].

> 'Procurement and availability is a big one generally, we have tight timeframes when we manufacture, generally. We have production programming; we generally have tight production programmes. I guess material availability would be a big one' [C_1P_3].

> 'Because, if you're already established in the market in terms of like BlueScope selling the coil and selling the purlin sections and everybody is used to that, there's no motivation for a disruptor in the market, just because the market is so small' [C_1P_1].

> 'There's not the ability for them [ATCO] to say, "I've got two products, and I can choose a recycled one, or high recycled content or a low recycled content one," in terms of the Australian landscape for steel products, what you get is what you get' [C_1P_6].

6.4.6. Lack of Organisational or Regulatory Targets for using PwRC

Organisational or regulatory targets can significantly encourage the use of PwRC by providing clear goals and incentives for organisations. These targets provide a structured approach to incorporating PwRC, aligning organisational practices with broader environmental goals. The lack of established targets for using these resources can reduce the motivation, accountability, and innovation needed to effectively incorporate them into construction projects.

The findings from the interviews indicate that, at present, there are no targets for using PwRC in modular construction within the study organisation, nor is there any policy in place to encourage their use. Some relevant responses are provided below:

But at this point, we're not yet into that area. We're not yet sure, we're still in like the reviewing process stage of those products and like assessment, but it hasn't been making up a firm call on those alternative reusable materials [C_1P_2].

We don't tend to use a lot of recycled products if I'm being honest. Then, when we do, it would usually be in line with the tender requirements [C_1P_3].

'We might be procuring something that has raw waste blended into it. But that's the same thing, we're not necessarily looking or setting expectations to our suppliers that, "Hey, we're going to procure this, and we want it to have at least 5% blended waste product in it because it's a society, a sustainability initiative." We haven't got to that point' [C_1P_5].

'I guess the government is all about being green as well…[but] they don't currently have any policy around using recycled material, they don't tend to' [C_1P_3].

'In regional areas or lower tier builders who don't have those things [obligatory incorporation of recycled content] commercially put into their deals, you know, they don't need to get recycled product. It's not mandated for them to do that. Therefore, they are often the smaller builders who will take the cheaper option' [C_1P_4].

6.4.7. Lack of Precedent in PwRC Application

One participant observed that the modular construction industry tends to be cautious about using products that lack market testing. Demonstrated performance from previous use can build market confidence and foster wider adoption of these products. C_1P_2 noted that while laboratory tests are a useful way to understand the nature of a product, they do not equate to assessing its actual utilisation and performance in modular components.

'It's more like a salesperson walking in, "Hey, we've got this recycle substrate board," right? It hasn't been tested on the market and it's more like they want us to have a look at that. So, we're kind of looking to other what other competitors are doing at the moment where the other competitors are' [C_1P_2].

'Like I said, it's been tested in the laboratory, but when it comes to if it's in construction itself, it hasn't been tested in that area' [C_1P_2].

'Whereas in the industry they introduce like different recycled materials. But then when we use those recycled materials in apartments, they can catch fire and it's the problem is that there are a lot of other companies that produce equivalent to those materials' [C_1P_3].

6.4.8. Health and Safety Issues

The use of PwRC presents health and safety complications. Physical and chemical changes during recycling processes can cause the industry to be cautious about using recycled materials due to potential health and safety issues. C_1P_2 highlighted that it is not advisable to use PwRC which lacks accompanying safety data.

'One of the issues that come from potentially looking at recyclable materials can come from, how good the data is around the safety of that product, and therefore the understanding and interpretation of potential controls or exposure for the workers. But do we know if it's safe to use when we're cutting it and sending it? Heating it? Glueing it, etc.? Washing it? That probably provides a little bit of fear response, which prevents them from looking at that space at the moment' [C_1P_5].

6.4.9. Lack of Certification for Recycled Products

As indicated in the previous category, the unknown nature of recycled materials can hinder their widespread application in modular construction. The lack of a formal assessment of PwRC performance, safety and quality can lead to uncertainty and reluctance in their adoption, as stakeholders may have insufficient information to evaluate their suitability and reliability for construction projects.

'Only a few hundred dollars later, I can get a test done and have it put to them [material supplier] that it's actually not what you say it is' [C_1P_1].

'The problem is that there are a lot of other companies that produce equivalent to those materials but haven't been tested or going through the process of assessing the nature of those products. And that's why we have the issue in the industry of materials that get caught in a fire and there was no proper certification' [C_1P_2].

> '…but what about the safety data sheet? What about the health?" and the person that was providing it to us didn't have the greatest knowledge yet… So, because those things weren't answerable, we had to say, well, we can't use the product, but to me it looked like a great product' [C_1P_5].

6.4.10. PwRC Longevity and Performance

In the building and construction sector, the longevity and performance of materials are crucial for the successful delivery of projects. Uncertainty regarding the longevity and performance of PwRC has been cited as a major factor limiting their use in modular construction. Several participants emphasised the importance of addressing this concern when considering the procurement of PwRC.

> 'But the cost and availability of this product and performance of the product is, is one of the key areas that we can consider at this point' [C_1P_2].

> 'I think businesses would be open to using recycled materials as long as there was longevity in the material' [C_1P_3].

> 'So, there's a bit of a market expectation on us as well to make sure that what we're providing is of a higher or highest quality' [C_1P_4].

6.4.11. Lack of Experience Dealing with the PwRC Supply Chain

As previously noted, the study organisation currently has limited capacity to incorporate PwRC into its modular products. One participant reported that the organisation does not have a procurement strategy specific to PwRC. Consequently, their ability to effectively engage with the PwRC supply chain is also constrained. Two participants observed that this limitation could affect their procurement processes, even if the organisation is willing to enhance its sustainability impacts through the use of PwRC.

> 'So that's 100% something that I think ATCO is aware that it's got to get, but I don't know if ATCO has enough experience yet to be able to understand how to have those conversations with suppliers to then target that and look at that [C_1P_5].

'The process of procuring waste materials would be no different to procuring any material. We would be, you know, trying to find it, we would be trying to understand it' [C_1P_5].

'I think, it's really the challenges that they face around increasing recycled content is around engaging those suppliers and what they're doing to drive, you know, drive change in their manufacturing route and yeah, so for us, it's something we've solved, but for those other manufacturers that I spoke about they're, they're each on a journey, those journeys are on different trajectories' [C_1P_6].

6.4.12. Reliance on Mainstream Supply Chain

The study organisation predominantly depends on mainstream, or shorter supply chains- as described by C_1P_1- for material procurement. Although this procurement strategy offers practical benefits, such as greater control, it can significantly hinder the adoption of PwRC. Indeed, this limited pool of suppliers may struggle to provide a diverse range of PwRC that meet the requirements.

'We use mainstream suppliers, Armstrongs, the Tarketts, the Polyflors and those kinds of things. So, if a fourth player were to enter into the market, maybe a smaller player, but maybe had a higher recycled content, and even if they were cost competitive, that would be a challenge to replace that in, because it becomes a company decision and rather than a project-specific decision if you like' [C_1P_1].

'So what I mean by a shorter supply chain is, if we have an established supplier who holds stock, who we deal with very well, we already have the correct payment terms and all that other kind of stuff' [C_1P_6].

'In terms of the materials purchased and procured to manufacture new products and to maintain old products, what I found with ATCO is that they've got very long-standing relationships with suppliers that they've managed over years and years and years and moving away from those - and they've got great cost control because of that - and investigating new is where they haven't really had procurement focus' [C_1P_5].

'So, you know, that is the main challenge for them around accessing the required spread of materials that they would use from suppliers who have high recycled contents, and they don't have the ability to switch those out' [C_1P_6].

6.4.13. Varying PwRC Specs in Modular Construction

It is a fact that sometimes the application requirements vary between modular and standard construction. As such, the PwRC suppliers need to meet varying requirements for the same material if they want to be engaged in both modular and standard construction markets. Two participants raised their concerns about this issue:

> 'Determining its fit for purpose, particularly within the module of building space where sometimes the parameters extend beyond, what we'd say, traditional building materials where I've got to worry about things like. So, things like say calcium silicate board is very good in traditional buildings, but you wouldn't use it in a modular building because it would crack. It would break unless you applied a certain glue, a certain consistency, and having, you know, you would have to alter the engineering parameters about the use of the product just to incorporate that within a modular structure' [C_1P_1]

> 'Like the majority of all these sales reps, "Yes, it meets the requirement of the NCC [National Construction Code]." But when it comes to construction then, if we applied install in a separate way different way, it's not going to meet the requirements, or it was going to lose the warranty of those products' [C_1P_2].

6.4.14. Limited Reporting Capacity

Both the organisation and the supply chain lack a mature infrastructure for reporting on PwRC usage. The absence of data needed for planning and decision-making in modular construction can discourage key stakeholders, including sustainability-focused clients, from using PwRC in projects.

> 'So sometimes we end up finding in construction, we're ticking a box without actually having any data that's usable from that, and the industry's not there necessarily learning' [C_1P_5].

> 'We've got limited data on the used side as opposed to the purchase side, and we're trying to get on top of that' [C_1P_5].

> 'How are we, what like, what recycling products are we using? How am I reporting through NGER [National Greenhouse and Energy Reporting] my impact on that project? So, what ends up happening is they go, "Look, we still have to get you to fill it out. So just put it, just send this e-mail with the reporting template every month with zero' [C_1P_5].

> 'The transparency of waste diversion ratios is the harder part because each individual supplier of those waste services will have different and varied levels of ability to report for us. So that's the bit that I've struggled with in the past two years, is getting consistent data to then analyse that data and set targets for improvement on diversion ratios for waste' [C_1P_5].

6.4.15. Limited Capacity to Integrate PwRC into Production Processes

Despite the organization having its own manufacturing unit, its current business model does not promote the internal integration of recycled materials into modular components. According to C_1P_5, the organisation typically purchases final products that may or may not include recycled content, a practice that appears to be standard across the industry. This approach significantly limits the potential to optimise the use of PwRC in modular construction, as it hinders the organisation's ability to directly influence and incorporate sustainable materials into its production processes.

> '[We use recycled materials] If the base material comes with that built-in, so one of the key things that I think it'll come in and one of the later questions, is that we use a such a small palette of materials' [C_1P_1].

> 'Getting that as a raw product is not gonna happen for ATCO 'cause we don't necessarily use a raw product. We then use, you know, we use steel, we don't get the raw product and then blend it with steel because we don't make our own steel. We don't make our own wall panelling etc. So, we're always using the end product. So, we would never be directly procuring raw waste' [C_1P_5].

6.4.16. Short-Term Thinking in the Decision-Making Process

Short-term thinking in decision-making can restrict the organisation's ability to embrace innovation and change, especially in areas like the circular economy in modular construction. Given that the benefits of sustainability initiatives often accrue over the long term, transforming decision-making processes to prioritise long-term outcomes is crucial. Some participants' responses on this issue are provided below:

> 'I guess one of the challenges with short-term thinking is that there's a cost involved to it... You know, if sustainability is #1, then yeah, we're looking great. However, if the price is #1, then those options can look quite expensive in the short term' [C_1P_4].

> 'The issue is that we haven't made up a decision yet because it's more like a salesperson walk in... we're kind of looking to other what other competitors are doing at the moment where the other competitors are' [C_1P_2].

6.5. Drivers of Using PwRC in Modular Construction

The analysis of interviews showed that there could be six major drivers for using PwRC in modular construction. These factors are illustrated in Figure 11.

6.5.1. Supportive Organisational Targets

The incorporation of PwRC into modular products can be influenced by organisational targets/key performance indicators (KPIs). These targets catalyse the required change in the organisational procurement model.

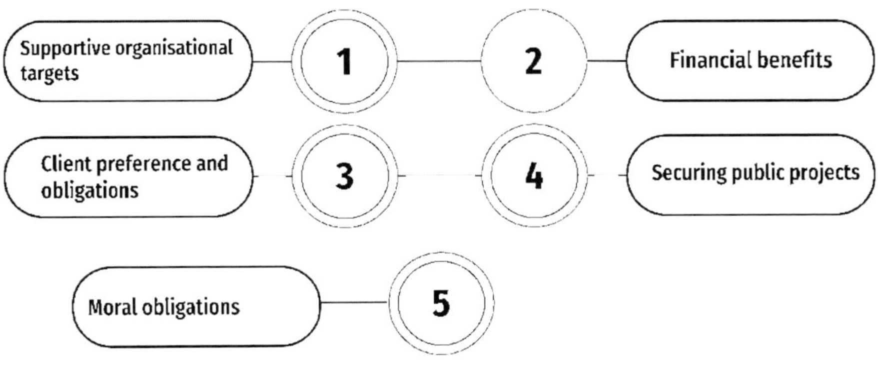

Figure 11. Major drivers for using PwRC in the case study.

> 'We've got a corporate sort of structure and sometimes, shall we say, mandates certain targets. So some of those targets get driven down into obviously managing directors and you know the director of the various business groups...so some of that decision-making can be KPI-driven in terms of we must now start to use this, if that's the directive, you know, we must reduce waste by this amount or increase utilisation of renewables by this amount and those kind of things' [C_1P_2].

> 'The main motivators for us, though, our opportunities on the supply side to, you know increase circularity to drive that economic and social return through diversion from landfill, the reduced greenhouse gas emissions' [C_1P_6].

6.5.2. Financial Benefits

The other motivating factor relates to whether these materials generate any financial benefits. According to the interviews (*C_1P_2, C_1P_2 and C_1P_4*), this factor can be a significant change facilitator in the decision-making process. However, these criteria vary depending on the material. For instance, the recycled alternative in steel is generally more cost-effective than in timber products. Furthermore, the long-term financial benefits of utilising these resources can be realised by integrating the reuse of waste materials into standard business practices. Over time, this will lead to a reduction in production and operational costs *(C_1P_4)*.

> 'So, the decision-making, particularly when it comes to using products with recycled content it either typically driven by a new product being profitable to the company' [C_1P_2].

> 'If you take a long-term stance as the business, quite often, a recycled product makes financial sense in the long term... Over time, it is actually a lot cheaper because there is no continued wastage and labour involved in redoing the same thing over and over again' [C_1P_4].

6.5.3. Client's Preference and Obligations

The third factor relates to the client's preference for using PwRC. This inclination may stem from their need to showcase their sustainability commitments, whether by reporting to public agencies, aiming for formal recognition through a sustainability rating system, or publicly promoting their dedication to sustainability. The client's preference typically is translated into contractual obligations. In this space, the government plays a key role in pushing a sustainability agenda including using PwRC in their assets and construction projects.

> 'Some of our, shall we say, bigger clients, particularly in the mining and resource sector, they have global minimum standards, and they want full material traceability and that kind of things, and that pushes you down a certain supply chain, if you like.' [C_1P_1]

> 'Basically, we received a lot of requests from education tender...But yeah, predominantly the education sector across Australia is the driving force behind introducing new recycled materials. It's all about that, helping the environment and things like that [C_1P_2]

> 'Really it comes down to contractual obligations, we rarely find a cheaper cost alternative with recycled products, I guess in application to our buildings I would say' [C_1P_3].

> 'Whereas tier one builders they, you know, report to the government, they're doing government work so they are held commercially to these standards...like a Green Star checklist that they would have to fill out, which says, "how do you do this? How do you do that? "In which case, they would need to engage sustainable materials or practises to be able to tick those boxes' [C_1P_4].

> 'So, what it is that a customer needs as part of their contractual requirements or obligations from their customers as well... contractual obligations generally, or often, on government projects, be they transport infrastructure projects that drive requirements for very high recycled contents across all materials' [C_1P_6].

6.5.4. Securing Public Projects

In some cases, the client for ATCO S&L products is the government which is typically in favour of enhancing sustainability in their procurement (C_1P_4). The growing demand for environmentally sustainable practices, including the use of green materials in government construction projects, is a significant driving force behind the adoption of PwRC in modular construction.

> 'We have received a couple of requests [tender documentation] from especially education, and they introduce, and I've seen a couple of new

materials being introduced to get on to it, to get a quote or design based off those materials' [C_1P_2].

'It definitely ticks a box for a lot of government jobs. So, we need to be playing in that space because otherwise, you know, if we're not being sustainable, we're not using recycled materials, it could be the difference between us winning and losing a job. So, we want to make sure that we're putting ourselves in the best position to win these jobs as possible' [C_1P_4].

'I would call it a regulatory role, in terms of the major rating tools and a regulatory role in terms of any particular government or other organisations that provide or drive the sustainability requirements… transport infrastructure projects that drive requirements for very high recycled contents across all materials' [C_1P_6].

6.5.5. Moral Obligations

The final factor pertains to the moral imperatives and ethical considerations that drive individuals to adopt sustainable practices. According to three participants (C_1P_1, C_1P_4 and C_1P_5) in the organisation, staff would like to push the sustainability agenda:

we do still have a personal obligation as human beings to try and do our best to make sure that we're not overusing things that we don't need to. So, there's a moral obligation there' [C_1P_4].

'It's just business and there is an ethical aspect that you need to consider when you're, making those decisions, it can never be just business' [C_1P_1].

'Like any person, we know that as a society and as the population continues to increase and we continue to use more and more of the raw natural resources of Earth, we're going to have to get better at recycling and we're going to have to get better at reusing recycling' [C_1P_5].

6.6. Strategies to Optimise the Use of PwRC in Modular Construction

The next section of the interview focused on strategies to effectively remove or minimise the challenges associated with utilising PwRC in modular construction. The analysis of interviews identified 10 major strategies that companies involved in modular construction in Australia can employ to increase the uptake of PwRC in their products. These strategies fall into three key domains: organisational capacity building, supply chain management assessment and improvement, and sustainable business decision-making. Among these, the categories of "attitudinal changes and awareness raising" and "internal opportunities for material recyclability and reuse" were found to have the most significant positive impact. The identified categories are illustrated in Figure 12.

6.6.1. Long Term Thinking

The long-term thinking approach involves planning and deciding with a focus on future outcomes rather than just immediate benefits. At present, the use of PwRC does not necessarily yield immediate economic benefits, with both potential and tangible advantages often materialising only over the long term. Therefore, organisations in the modular construction sector aiming to enhance sustainability and achieve their business objectives must adopt long-term thinking in their decision-making processes, particularly when it comes to incorporating PwRC into their products. This point was made by C_1P_4 who suggested that long-term thinking is the key to success:

> 'These days, a lot of businesses are making a lot more, especially ATCO, long-term decisions as opposed to short-term decisions within the business to ensure that we're doing the right things to put us in a good frame coming into the next five, 10, 20 years' [C_1P_4].

> 'In the first instance, it [PwRC] is more expensive than what we're currently doing. However, over time, it is actually a lot cheaper because there is not the continued wastage and labour involved in redoing the same thing over and over again … so there was actually a business motive there to, you know, increase our profits, I guess, at the end of the day, over the long term' [C_1P_4].

> 'The economic, I think we've talked about that, which is the fact that a lot of these recycled products have a longer life span than the other ones. So, whilst they may be a higher initial investment, they certainly provide better economic advantages over the long term' [C_1P_4].

6.6.2. Historic Application of PwRC and Confidence-Building

The successful application of PwRC across the industry can significantly boost confidence in utilising these resources. Two participants in this study emphasised that their organisation would be more likely to adopt PwRC in their products if these materials had been used in real-life projects and relevant data were available. Creating demonstration modular projects where the study organization successfully integrates PwRC offers a unique opportunity to lead the market and set a precedent in this area.

> 'If this product's being used, if there's a positive historical background of this product being used by the industry then, and it's cost effective, then the company will go down into that line' [C_1P_2].

> 'for ATCO in particular, to move into looking further at recyclable materials in their manufacturing space, one of the things that would make it easier is having that information available to be more readily presented rather than having to do their own digging and investigation' [C_1P_5].

Source: Authors.

Figure 12. Strategies to enhance the use of PwRC in modular construction.

6.6.3. Recycled Product Information and Certification

Three participants noted that certified information on PwRC can inform their procurement process. However, the effectiveness of this process can be hindered by the existence of various certification schemes and different product information. To address this, the organisation should develop new or utilise existing standard templates to ensure consistent and reliable product information is received. In this process, consulting with trusted third-party PwRC certifiers and major suppliers is essential. Such a template should outline the product's features (such as physical and chemical characteristics) as well as its functions (including quality, longevity, and performance). The template must align with sustainability reporting requirements, including Green Star, IS rating, Leadership in Energy and Environmental Design (LEED), and NGER. According to C_1P_2, the product information should be prepared and produced to demonstrate their conformity with NCC.

> 'So, whether they [PwRC] are fit for purpose and those kinds of things, so I usually get when a salesman comes in. I don't necessarily speak to the salesman, but I will speak to the data that they provide or don't provide' [C_1P_1].

> 'Some sort of reuse of recycling materials and it should come with evidence, producing evidence that these products have been tested on the market and there's some companies, that they're doing all series of testing and certificates of conformance that it meets the requirements of the NCC' [C_1P_2].

> 'It's very easy to get the upfront cost analysis, it should be easy [to use PwRC] if the provider of a recycled product has all that information around their product' [C_1P_5].

6.6.4. Effective Marketing

Being able to communicate the sustainability practice of the organisation can offer a series of benefits. According to C_1P_5, these benefits may include having a positive return on the business and more clients would like to work with the organisation compared to their competitors. Sustainability is a key selling point for businesses in the current and future industry climates. The use of these products in modular construction can be aligned with many sustainability initiatives including Environmental, Social, and Governance (ESG) and Sustainable Development Goals (SDGs); Reporting the achievement of these goals, along with promoting the improved organisation's socio-environmental profile through targeted marketing tools, can enhance visibility across industry, government sectors, and the general public. This increased visibility can lead to greater customer reach, stronger brand recognition, and enhanced credibility within the market.

Effective marketing strategies are essential for achieving the best outcomes. These strategies should communicate and highlight the journey behind the use of PwRC, the organisation's achievements, and the benefits they offer, ensuring maximum visibility and impact.

> 'You can offset cost for cost, if we can then also demonstrate how we can market it and ATCO's not a big marketer, a self-marketer in the industry. We need to, you know, one of my tasks [...] is to drive more marketing. Even if that's B2B marketing for ATCO structures in Australia, one of those things is we need to get better at marketing and that's where having improved ESG is a marketable thing for every business in Australia moving forward' [C_1P_5].

> 'To be able to lead it, from a marketing perspective, from a global perspective and show how good we can be in this, what prevents me from doing that, it's a really good business case to say we can produce buildings at the same rate and the same volume that we're doing, so with these products, these products are better for the environment because they are recyclable' [C_1P_5].

> 'We can present that; we can be proud of that and that's going to actually have a positive return on our business because more and more people will want to work with ATCO rather than with one of our competitors. So, I, yeah, absolutely look at it and we should be kicking the door down in this space' [C_1P_5].

6.6.5. Attitudinal Changes and Awareness-Raising

Several challenges highlighted in Figure 10 can be effectively addressed with this strategy. In particular, it can help overcome the industry's conservative approach and short-term thinking in decision-making. By raising awareness among key stakeholders about the potential and actual benefits of using PwRC, this strategy can accelerate the procurement and application of these products in modular construction projects.

> 'But I think if, everyone understands the nature and the concept behind the using of recycled materials, I think, it can change the methodology, it can change the mindset, especially who's driving the construction' [C_1P_2].

> 'Yeah, I think probably increased knowledge and awareness around particular products would be key. Understanding how it would benefit our processes and our methodology construction methodology in modular. I think largely people would be happy to entertain using recycled products as long as we understood the benefits, better understand' [C_1P_3].

> 'It's more getting the big players in the modular industry, like we said earlier, to change their mindset so that they can accept like for like, or not like for like, but like for different with a better outcome' [C_1P_5].

> 'It's being able to present all the answers to them [senior managers in modular construction companies] before you even ask them to look at it, is the key driver for success for me' [C_1P_5].

6.6.6. Leveraging the Purchasing Power of Key Players

Key players in the modular construction industry, such as ATCO S&L, possess significant influence to make the use of PwRC a standard practice across the sector. As market leaders, these companies should focus on building internal capacity to effectively procure and integrate PwRC into their products. Once these practices are established as the industry norm, it will encourage SMEs to adopt similar procurement strategies, further embedding PwRC use in the industry. According to C_1P_4, their purchasing power can generate sustainable demand for these products, potentially leading to economies of scale and a reduction in production costs. The organisation and the recycling industry can capitalise on the advantages of mass production, leading to enhanced profit margins.

> 'If the big companies are not pushing and being part of the strategy, then it's never going to work. It's never going to work [C_1P_3].

> 'ATCO's a big company, we have a lot of buying power. You know, we could start to look at creating partnerships with companies who sell the products that we require for sustainable or recycled materials so that we

can look to get better buying power and therefore create a lower price point for our clients' [C_1P_4].

6.6.7. Setting Targets Through Organisational or Government Policies

Setting targets for using recycled content in products can guide processes involved in the procurement and application of PwRC in construction projects. Reasonably set targets can drive numerous changes within organizations, including fostering innovation, promoting accountability, building market demand, and encouraging collaboration across material supply chains. According to C_1P_3, the government can play a pivotal role in setting these targets by developing policies such as sustainable procurement guidelines.

> 'We've got a corporate sort of structure and sometimes that [...] mandates certain targets. So, some of those targets get driven down into obviously managing directors [...]. So, some of that decision making can be KPI-driven in terms of we must now start to use' [C_1P_1].

> 'I think it would help if there was a push, potentially from the government as well...They don't currently have any policy around using recycled material, they don't tend to' [C_1P_3].

> 'Both from an initial, you know, what are we doing right now? Where's my benchmark? Where's my baseline in that? And then to be able to sit there and go, OK, there's a massive opportunity to lead in this space, rather than to follow' [C_1P_5].

6.6.8. Internal Opportunities for Material Recyclability and Reuse

Organisations in modular construction can plan for maximum recyclability and reuse by incorporating these considerations throughout their manufacturing, operation and execution phases. Such incorporation can be more effectively implemented through adopting a circular economy business model. This way they can influence their internal stakeholders to advance the circularity of their products. Almost all participants in this study emphasised the importance of exploring opportunities to capture the value of waste resources and enhance the utilisation of PwRC in modular construction. For example, C_1P_1 and C_1P_4 stressed the need to revisit their production line to evaluate opportunities for material recyclability and reuse. Similarly, C_1P_6 noted that their organisation aims to recapture product waste resources for repurposing and reuse through extended producer responsibility practices.

'Simple builds are able to be reused and we try, we sell on that as well. We sell on the fact that we can actually bring those back into the mainstream with very little in the way of upgrades and most of those are cosmetic and those kinds of things' [C_1P_1].

'So, what we try to do there as much as possible and where possible is reuse certain materials. So, for example, our decks, we have steel framed decks…So we will take that part off the deck. We will throw that part out. But we will reuse the deck frame, reline it so that we don't have to continually go and get more steel framing built for our jobs' [C_1P_4].

'But I think ATCO is also going to try and match that wherever we can and where it makes sense in our product lines so that we're providing it as a standard as opposed to on a job-by-job basis' [C_1P_4].

'Now, they [low-grade polyply wall linings]'re great. They're cheap. They go up. If a client damages it, you just unscrew it, you put a new sheet up. So being able to look at recyclable materials of that and in effect, it already is recyclable to an extent' [C_1P_5].

'it's not so much the need for modular to look at how do we generate a product in the first place? It's what products are out there, and how do we use them in our existing processes? It's not that the products and the research aren't already out in the market for recyclable, it's, "OK, where can we use this in the modular game" [C1P5].

'So as a business that is very much our focus from a point of view of maximising recycled content and then the idea of keeping that in, through its use phase and then to its end-of-life phase and then recapturing that back through our recycling business or indeed any of the other recycling channels or reuse opportunities to bring that back into as a new product or as a reused product to keep that loop going' [C_1P_6].

6.6.9. Active Engagement in the Material Supply Chain

Modular construction companies should actively engage in the material supply chain to influence their external stakeholders to achieve material circularity outcomes. Changes within the supply chain should focus on waste reduction, increased utilisation of recycled content, and the enhancement of PwRC quality, longevity, and performance. According to C_1P_4, these changes should stem from strategic partnerships with material suppliers involved in PwRC production and supply.

> 'Re-recycle it into our own product, if you like. So, if we had a closed-loop supply chain in terms of, say, the vinyl manufacturer…if you had a supplier of material down the road where, you know, the off-cuts could be taken straight back, put in a hopper, chopped up and then used to produce the same material again' [C_1P_1].

> 'So, get me stillages that can be reused if they, if a client says they can't. Or a supplier says they can't supply it in the correct packaging to be reused. Then you send the packaging back with the next truck when they deliver' [C_1P_1].

> 'You know, we could start to look at creating partnerships with companies who sell the products that we require for sustainable or recycled materials so that we can look to get better buying power and therefore create a lower price point for our clients' [C_1P_4].

> 'Is there other things that you can blend into it like I was talking about? You know, microplastics. Can we get microplastics? … Is there a potential way to actually blend those plastics from a container bottle scheme into the timber product with the bonding agents so that you've got a semi-waterproof product that then has a higher lifespan, and therefore we're not replacing it as frequently' [C_1P_5].

6.6.10. Products Buyback Agreements

The final strategy involves creating contracts that explicitly stipulate a product buyback agreement between modular manufacturers and end-users. Given that modular products are not always used permanently, a buyback arrangement provides an opportunity for manufacturers to repurchase these items from end-users. According to C_1P_1, this agreement offers mutual benefits: end-users can generate revenue by selling unused products, while manufacturers gain the opportunity to recover modular products either fully or partially. The materials salvaged from these end-of-life modular products can then be repurposed in new modular builds, promoting sustainability and resource efficiency.

> 'So sometimes we will sell a building with a buyback in there because we are better able to reuse and repurpose that building. The clients might have it for 5- 10 years and it's not economical for them to rent it for that

period of time, but it is economical for them to buy it, and then we will buy it back from them for an agreed price' [C_1P_1].

6.7. The Future of PwRC Applications in Modular Construction

The interviewees suggested that a broader application of PwRC in the modular construction industry is plausible in the future. However, several participants emphasised the necessity of specific changes for this to materialise. These changes include overcoming the sector's conservative approach, its mindset transformation, preparing for a transitional phase, and achieving greater adoption within the conventional construction sector. Table 4 presents the key points highlighted by participants in response to this question.

Table 1. Participants' views regarding the future of PwRC applications in modular construction

Participant	Key point	Quote
C_1P_1	Challenges posed by sector conservatism	*I think the future applications of recycled materials will be in new product development because the construction industry, by its nature, is very conservative. But, like I said, a lot of things are based on, "It worked last time. I will do it again." And so, you always end up with an 'It's almost like a generational change'*
C_1P_2	Initially complex, ultimately ubiquitous	*I think it can be complicated for a start, but then for a certain period of time in that transition period, I think everyone will start picking up and say, "OK, yeah, it's worth going with recycle*
C_1P_3	Mainstream upon sector's adoption	*If there's a place for recycled material in construction, there's probably no problem in a ship to modular construction to utilise those projects, those products*
C_1P_4	Expanding with growing PwRC development	*I think it's only gonna get bigger and bigger, personally. More and more products are being delivered and created with recycled materials*
C_1P_5	Dependent on the industry mindset transformation	*It's not so much the need for modular to look at how we generate a product in the first place? It's what products are out there, and how do we use them in our existing processes? It's more getting the big players in the modular industry... to change their mindset so that they can accept [...]like for different with a better outcome.*
C_1P_6	Growth driven by urban development needs	*As countries and as cities develop, that supply chain of recycled material increases because you put it in once and as you redevelop, rip down, repair, rebuild, that steel comes back out and into the recycling channel*

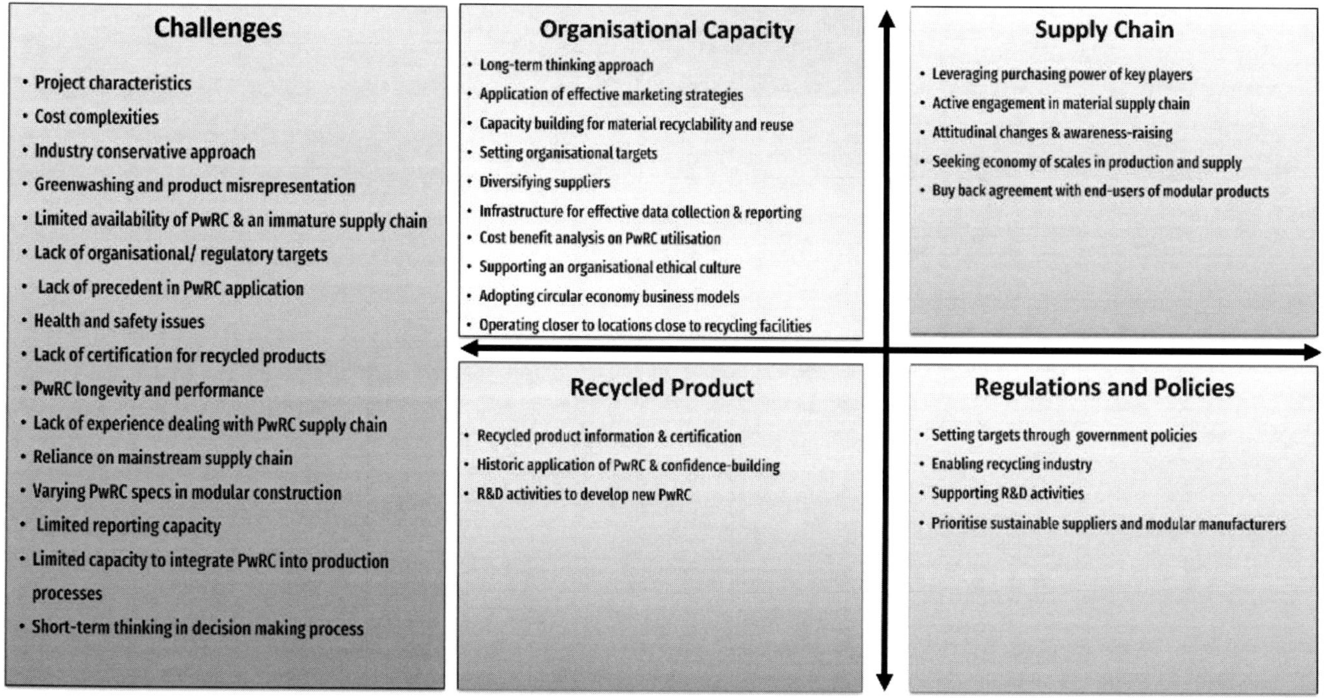

Source: Authors

Figure 13. A framework to influence stakeholders for optimal uptake of PwRC in modular construction.

6.8. A Framework to Enhance PwRC Uptake in Modular Construction

The following model is created based on the findings' literature review presented earlier and the insights captured in the interviews. The framework guides efforts in removing the identified obstacles to sustainable and optimal uptake of PwRC in modular construction. As outlined in Figure 13, the framework consists of organisational capacity, supply chain, recycled products and regulations and policies major components that address various obstacles identified in this study.

Conclusion

ATCO S&L may consider adopting the strategies outlined in the framework to enhance its capacity for increasing the uptake of PwRC in its offerings. The organisation can also lead and influence the supply chain by implementing the tools and strategies proposed in this study, making the use of PwRC a standard practice. To achieve this, the organisation should plan for a gradual implementation of these strategies, aligning them with their priorities and business objectives.

Currently, many Australian governments are under pressure to meet the requirements of the SDGs and various climate change initiatives, such as the Paris Agreement. Government agencies involved in the sustainability of built environments and construction activities can leverage the findings of this study to evaluate, plan, and prioritize the most effective actions to enhance sustainability outcomes moving forward.

The building and construction sector should pay close attention to the modular construction space, as it is increasingly being proposed as a viable solution to housing crises both in Australia and globally. The challenges identified in this study can help the industry prepare for a transition towards more modular and sustainable buildings in the near and distant future. Thus, the sector has an excellent opportunity to reduce its environmental impact by incorporating PwRC in modular building products.

References

[1] ATCO Structures and Logistics. *Off-site modular construction offers efficiencies over in-site*. 2024 [cited 2024; Available from: https://bit.ly/3zoTAFn.

[2] KPMG, *Smart Construction – How offsite manufacturing can transform our industry*. 2016.

[3] Navaratnam, S., Satheeskumar, A., Zhang, G., Nguyen, K., Venkatesan, S. and Poologanathan, K., The challenges confronting the growth of sustainable prefabricated building construction in Australia: Construction industry views. *Journal of Building Engineering*, 2022. 48: p. 103935.

[4] Bertram, N., Fuchs , S., Mischke, J., Plater, R., Strube, G. and Woetzel, J., *Modern Construction: From Projects to Products.* . 2019, McKinsey & Company.

[5] prefabAUS, *Australia's Smart Building Revolution: A Prefabrication Industry Roadmap 2023-2033*. 2023.

[6] Lawson, R. M., Ogden, R. G. and Bergin, R., Application of modular construction in high-rise buildings. *Journal of Architectural Engineering*, 2012. 18(2): p. 148-154.

[7] Meyers, M., *How modular construction is keeping waste out of U.S. landfills. 3P Contributor*, in *Triple Pundit*. 2016.

[8] Loizou, L., Barati, K., Shen, X. and Li, B., Quantifying advantages of modular construction: Waste generation. *Buildings*, 2021. 11(12): p. 622.

[9] Latham, D., *The future is prefabricated: Prefabrication and modular construction*, in *Latham*. 2024.

[10] IBISWorld, *Industry Report: Prefabricated metal building manufacturing in Australia*. 2023.

[11] Rose, T., *Emergency housing rollout across flood-hit NSW north coast a 'dog's breakfast,' critics say*, in *The Guardian*. 2022.

[12] Rose, T., *Buyback offers to reach 1,100 northern NSW homeowners, 500 fewer than hoped by agency boss*, in *The Gurdian*. 2023.

[13] Minunno, R., O'Grady, T., Morrison, G. M. and Gruner, R. L., Exploring environmental benefits of reuse and recycle practices: A circular economy case study of a modular building. *Resources, Conservation and Recycling*, 2020. 160: p. 104855.

[14] Hammad, A. W., Akbarnezhad, A., Wu, P., Wang, X. and Haddad, A., Building information modelling-based framework to contrast conventional and modular construction methods through selected sustainability factors. *Journal of Cleaner Production*, 2019. 228: p. 1264-1281.

[15] Shooshtarian, S., Maqsood, T., Wong, P. S., Caldera, S., Ryley, T., Zaman, A. and Ruiz, A., Circular economy in action: The application of products with recycled content in construction projects: A multiple case study approach. *Smart and Sustainable Built Environment*, 2024. 13(2): p. 370-394.

[16] Mendi, P., *Circular economy, modular construction and recycled materials*. 2022, Centre for Advanced Manufacturing of Prefabricated Housing: Institution of Civil Engineers Australasia.

[17] Shooshtarian, S., Maqsood, T., Wong, P. S. and Bettini, L., Application of sustainable procurement policy to improve the circularity of construction and demolition waste resources in Australia. *Materials Circular Economy*, 2022. 4(1): p. 1-22.

[18] Shooshtarian, S., Maqsood, T., Wong, P. S., Zaman, A., Caldera, S. and Ryley, T., Utilisation of certification schemes for recycled products in the Australian building and construction sector. *Business Strategy and the Environment*, 2024. 2024(33): p. 1759-1777.

Chapter 3

Local Collaboration in the Subiaco Oval Stadium Project

Salman Shooshtarian[1]
Peter SP Wong[1]
Tayyab Maqsood[1]
Atiq Zaman[2]
and Ana Maria Caceres Ruiz[2]
[1]School of Property, Construction and Project Management, RMIT University, Melbourne, Australia
[2]School of Design and the Built Environment, Curtin University, Perth, Australia

Abstract

This chapter offers an insight into the impact of local collaboration as an effective strategy in enhancing the economic, social and environmental performance of products with recycled content (PwRC) utilisation in construction projects. The content of this chapter is based on the review of the relevant existing literature and findings of interviewing the key stakeholders in the Subiaco Oval stadium project in Perth, Western Australia. The review of existing literature establishes the economic, social and environmental benefits that emerge from local collaboration on using PwRC in construction projects on a global scale. The findings from the case study illustrate the extent to which this strategy, which is materialised via using waste resources extracted from the demolition of this stadium in public infrastructure projects, could improve the environmental and economic performance of these resources. According to the findings, the main drivers for local collaboration to utilise PwRC in the case project can be categorised into three groups: 'government

In: Policy, Practice and Education of the Circular Economy in Built Environments
Editors: Salman Shooshtarian, Peter SP Wong and Tayyab Maqsood
ISBN: 979-8-89530-623-9

sustainability commitment and demonstration,' 'sustainability recognition at the project level' and 'policies and regulations.' The analytical results also show that 13 key challenges hindered the effective implementation of local collaboration for optimising the recovery of materials from the Stadium demolition waste and utilising the resulting PwRC in other construction projects. These challenges were then categorised under three groups: 'project physical characteristics,' 'project management,' 'supply chain and policies and regulations.' The research also provided 18 actionable recommendations that will enhance the use of PwRC through local collaboration. These strategies are classified into three groups: 'effective project development and management,' 'organisational capacity enhancement' and 'Government support and policies.' Lastly, the findings of this study suggest that local collaboration is emerging as a key trend in the use of PwRC, primarily due to various sustainability benefits.

Keywords: Subiaco oval stadium, products with recycled content, proximity principle, western Australia

1. Introduction

In recent years, the optimal use of products with recycled content (PwRC) has been a controversial topic in the field of construction sustainability. There are several strategies [2] that can potentially have a positive impact on the streamlined utilisation of these resources in construction projects. Local collaboration serves as a powerful strategy where various local stakeholders work together to utilise waste resources that are generated, recycled, and supplied within a certain region. This approach maximises the efficient use of local materials while fostering a circular economy at the regional level, reducing environmental impact and promoting sustainability. This strategy is also known as the territorialisation of the circular economy [3].

Several waste management policies support local collaboration to realise these benefits [4]. For instance, the proximity principle (PP) policy [5] requires organisations that generate waste to deliver it to a location for processing that is within a particular distance from where it originated. This circular economy policy has the potential to prevent the unnecessary transport of waste from one location to another, thus mitigating the need for and minimising levy liabilities while concurrently alleviating both the environmental and economic burdens associated with such practices. This policy is also described as the territorialisation of the circular economy [3].

Findings of a survey study showed that this policy, along with the manufacturer's shared responsibility for waste generation and subsidies for PwRC, were the top three waste strategies for the effective management of construction and demolition (C&D) waste materials [6].

Despite the potential benefits of local collaborations for using PwRC, there are several challenges along the way. These challenges can prevent or weaken forming collaborations in a region, promoting the stakeholders seeking alternative options.

2. Research Objectives

This chapter aimed to identify the key challenges and drivers for using PwRC in construction projects in Perth, Australia, by conducting a literature review and examining a case study of local collaboration. The three key objectives include:

1. To explore drivers for local collaboration in PwRC utilisation
2. To identify key challenges influencing the application of proximity principles in the building and construction sector
3. To determine key strategies to enhance local collaboration in using PwRC.

3. Research Approach

3.1. Review of Literature

This section of the report presents the research approach employed. First, the research team conducted an extensive literature review to offer insight into the role of local collaboration in the optimisation of using PwRC in construction projects. This review identified the main challenges, benefits and drivers of using this strategy in the building and construction sector. The sources used in this step included journal articles and government and industry reports that are freely available online. Care was taken to only capture the literature that describes the use of this strategy in recent years to reflect the contemporary efforts and advances in using these materials in construction projects.

3.2. Case Study

In the second step, a case study analysis was conducted to understand Australians' experience in using local collaboration strategy to enhance the use of PwRC in the building and construction sector. In consultation with the project partners and waste experts, the criteria for the selection of the case study were set. The case study had to meet the following criteria:

- Use of PwRC in a construction project
- Evidence of local collaboration for such a utilisation
- Ease of access to key stakeholders.

After identifying the project, Subiaco Oval Stadium (Perth, Australia), the perceptions of key stakeholder groups playing an essential role in using PwRC were sought through conducting a series of interviews. The interviewees were selected in consultation with the DevelopmentWA sustainability team. An invitation email was sent to each stakeholder to arrange suitable interview times.

Table 1. Interview questions used in the case study analysis

No	Question
1	Please introduce yourself, including your employment history and recent activities in the field of construction and demolition (C&D) waste management.
2	Can you please tell me a bit about the project and your respective involvement?
3	Can you please tell me about your organisation's experience using PwRC in this project?
4	What is the process of procuring waste materials and repurposing them into usable products in your organisation?
5	What were your main five motives for using recycled content in this project?
6	What were the main five challenges to using recycled content in this project?
7	In your opinion, how could these challenges (and their impact) be overcome?
8	What are the specific benefits and challenges of local collaborations for supplying and application of recycled materials?
9	What are the primary social, environmental, and economic advantages of using local collaboration in this project?
10	How do you envision the future of applying proximity principles to enhance the adoption of recycled materials in Australia?
11	Has your organisation published any relevant documents describing the project planning, design and execution that you could share with us?

The online interviews were conducted between March 2024 and April 2024 using the Microsoft Teams online platform. Each interview took a maximum of one hour and was recorded for transcription purposes. Three types of questions were asked during interviews: participants' details, their experience using the PwRC modular construction, and the enablers of and barriers to using such resources in this section of the industry. Table 1 provides the full interview questions.

3.3. Data Analysis

The data extracted from the literature review was thematically analysed preceding the creation of the major clusters of benefits, challenges and barriers to implementing local collaboration principles for optimal use of PwRC in construction projects. The literature analysis provided the basis for a case study analysis in the Australian building and construction sector.

Audio data captured from three interviewees were carefully transcribed by a professional transcriber using the word-for-word method prior to the quality verification of the text data by the research team. The transcripts were analysed using the NVivo Pro 12 application, which facilitates codifying text-based qualitative data. A thematic analysis method was applied to identify emerging themes that were related to the three research objectives. Furthermore, to better compare the qualitative data, quantitative descriptive analysis methods were adopted. The frequency of distribution of various categories of factors identified in the interviews was the main statistical measure used to conduct these comparisons.

4. Literature Review

There is a scarcity of evidence showing how local collaboration could result in the optimal utilisation of PwRC in construction projects. This section of the chapter explores how previous literature documented the economic and environmental benefits of utilisation of PwRC on a global scale.

Table 2. Environmental benefits of local collaboration in the utilisation of PwRC in construction projects

Key benefits	Description	Reference
Reduced carbon emissions	The waste is not transported over long distances by motor vehicles that produce carbon emissions	Halvorsen and Andersson [10], Wu, Zuo [11], Akhtar, Bani-Hani [12], Chembessi, Bourdin [3]
Improved waste data collection and reporting	When local governments, businesses, and organisations work together, they can establish standardised data collection methods, leading to more accurate and comprehensive reporting of waste generation, recycling rates, and material flows, ultimately enabling informed decision-making for environmental strategies.	Aslam, Huang [13], Almeida, Vieira [14]
Reduced pollution	Collaboration among local stakeholders, including waste generators, recyclers, and waste management authorities, facilitates the development of efficient transport routes and promotes resource sharing. This synergy significantly reduces overall pollution levels by minimising travel distances, optimising logistics, and decreasing fuel consumption.	Essex and Whelan [15], Halvorsen and Andersson [10], Wang, Luo [16], Jian, Wu [17]

4.1. Local Collaboration for Enhanced Environmental Impacts

Collaboration between waste generators, waste resource processors and the user of PwRC is highly recommended under the proximity principle. Challenges tend to emerge more prominently on smaller scales. For example, issues related to the proximity of facilities often surface, and changing behaviours typically necessitate collaboration among various stakeholders. In this regard, Ramos, Martinho [7] stated that it is crucial to understand that if local operations do not function effectively, achieving the broader goals of circularity within the building and construction sector is unlikely. This highlights the importance of local dynamics in the success of circular economy initiatives. When stakeholders work together effectively at a local level, it enhances the potential for reaching significant sustainability objectives in the construction sector.

In Australia, on the national level, in some states, such as New South Wales Proximity Principle is legally enforced for certain zones [8]. In this state, the maximum distance is 150 kilometres from the premises of the waste origin. Globally, the European Environment Agency has established this principle in the EU Framework Directive on waste [9]. Several environmental benefits are reported to be associated with meeting proximity principles when managing waste resources and utilisation of PwRC. Table 2 summarises these benefits.

First, the proximity principle reduces transport distances which in turn causes less carbon emissions. It is reported that PwRC can have lower embodied energy in addition to reduced transport emissions [12], especially where PwRC are used in close proximity to the site [2, 10]. Second, collaboration at the local level naturally fosters improved waste data collection and reporting [13, 14]. This enhanced data accuracy and availability, in turn, contribute to more effective environmental planning and decision-making within the geography of operation. Lastly, collaboration among local stakeholders can facilitate the optimisation of transport routes and encourage resource sharing - two key strategies that significantly contribute to reducing environmental pollution [10, 17].

4.2. Local Collaboration for Enhanced Economic Performance

As shown in Table 3, there are eight major economic benefits associated with using local collaboration to use PwRC. These benefits can significantly justify the use of PwRC in local construction projects. These benefits collectively highlight the economic advantages of local collaboration in promoting the use of PwRC in the construction sector, fostering sustainability while enhancing community resilience. The shorter transport distances can significantly improve the economic performance of local procurement of PwRC. The first apparent benefit of this is the reduction in transportation costs. Previous studies have reported that one of the major barriers to optimal utilisation of PwRC in construction projects is transportation costs [18, 19]. Several studies suggest that the use of PwRC resourced locally can significantly lower the total unit price of these resources, making them an economically attractive material for the construction sector [7, 12].

Furthermore, it is often argued that the production and utilisation of PwRC in close proximity not only alleviates pressure on road infrastructure [10] but also reduces the number of transportation vehicles required. The other two benefits relate to material handling and delivery. Theoretically, local procurement of PwRC minimises damage to materials [7] and reduces delivery timeframes. Local collaboration also supports the development and reinforcement of local supply chains targeting secondary construction materials [14, 15].

Table 3. Economic benefits of local collaboration in the use of PwRC in construction projects

Key benefits	Description	Reference
Reduced transportation costs	Sourcing PwRC locally decreases the distance they need to be transported, leading to significant savings on fuel and logistics costs. This not only lowers overall project expenses but also contributes to lower carbon emissions associated with transportation	Aslam, Huang [13], Treloar, Gupta [20], Ramos, Martinho [7], Akhtar, Bani-Hani [12]
Reduced complexity for managing waste resources	Local collaboration streamlines the management of waste resources, as stakeholders can coordinate more effectively to ensure proper handling, processing, and recycling of materials. This simplification can lead to improved efficiency and reduced administrative burdens	Ramos, Martinho [7], Almeida, Vieira [14]
Reduced pressure on the local and state road infrastructure	By decreasing the volume of transport trucks travelling long distances to dispose of waste or deliver materials, local recycling efforts can alleviate congestion and wear on road infrastructure. This can lead to lower maintenance costs and improved safety for local communities	Halvorsen and Andersson [10]
Reduced depreciation of transportation vehicles	With less frequent and shorter trips needed for transporting PwRC, the wear and tear on transportation vehicles is minimised. This can extend the lifespan of vehicles, reducing maintenance costs and overall depreciation	Authors
Increased local resilience	Collaborating with local recyclers and suppliers enhances the resilience of local economies by fostering self-sufficiency. Communities can rely on local resources for construction needs, reducing vulnerability to external supply chain disruptions	Walker, Redmond [21], Essex and Whelan [15]
Provision of sustainable feedstock for local waste recovery facilities	Local collaboration ensures a consistent supply of PwRC, providing a sustainable feedstock for local recycling and recovery facilities. This supports the circular economy and encourages investment in local waste management infrastructure	Almeida, Vieira [14]
Reduced lead time for material procurement	Local sourcing of PwRC can significantly shorten lead times, as materials are readily available within the community. This allows for more efficient project timelines and reduces delays associated with long-distance procurement	Authors
Reduced impacts on the material quality	Collaborating locally can enhance quality control measures for PwRC, as stakeholders are more closely involved in the sourcing and processing. This can help maintain high standards and ensure that PwRC meet the necessary specifications for construction	Ramos, Martinho [7]

4.3. Local Collaboration for Enhanced Social Performance

In addition to environmental and economic benefits, there are social benefits of local collaboration for optimal use of PwRC in proximity. Five social benefits associated with this strategy are identified and detailed in Table 4. The main social benefit is the growth in local job opportunities [15, 22]. Wu, Zuo [23] argued that the regional waste transfer will increase employment in the local waste recovery industry and cause significant positive social impacts on the communities involved in waste management activities.

Table 4. Social benefits of local collaboration in the utilisation of PwRC in construction projects

Key benefits	Description	Reference
Creation of local jobs	Initiatives focused on recycling and the use of local materials can create job opportunities in the recycling sector, construction, and related industries	Essex and Whelan [15], Yadav and Sinha [22], Wu, Zuo [11]
Community engagement and empowerment	Collaborating on recycling initiatives fosters a sense of community ownership and participation. Local stakeholders can work together towards common sustainability goals. This engagement can empower communities by encouraging active participation in decision-making processes and fostering a sense of pride in local initiatives	Walker, Redmond [21]. Wu, Zuo [23]
Enhanced social cohesion	Working together on environmental projects can strengthen relationships among community members and organisations. Such collaboration builds networks and trust, which can enhance social cohesion and create a more resilient community	Macedo, Huertas [24]
Improving local constructors' capacity to handle with PwRC	By working together, local construction firms can share knowledge and resources, improving their capacity to effectively utilise PwRC. This capability enhancement can lead to higher quality construction practices and more successful projects	Ramos, Martinho [7], Almeida, Vieira [14]
Educational and awareness opportunities	Local collaboration facilitates the sharing of knowledge and resources, improving awareness of recycling practices and sustainability issues among community members. Educational programs can empower individuals and organisations to make more informed choices regarding waste management and resource utilisation. This knowledge can inspire future generations to adopt environmentally friendly habits and practices	Macedo, Huertas [24], Hafizah Hammad Ahmad, Nabila [25]

5. Case Study Background

Subiaco Oval (also known as Domain Stadium), situated in the Subiaco suburb of Perth, was a sports stadium inaugurated in 1908. Its operations ceased in 2017 with the unveiling of the new Perth Stadium in Burswood. When it was in operation, the stadium held a capacity of 43,500 individuals, which was the largest stadium in Western Australia (WA) and a prominent sports venue in the country (Figure 1).

Source: DevelopmentWA [1].

Figure 1. Subiaco oval stadium.

Source: DevelopmentWA [1].

Figure 2. Overview of demolition activities in Subiaco Oval stadium project.

Source: Hocking Heritage+ Architecture [26].

Figure 3. Some of the features extracted from the stadium were sold at auction.

The demolition of the stadium was the largest demolition activity across the state. The demolition began after its final sporting event in November 2017. The demolition was part of a broader urban transformation of the area, preserving parts of the stadium's legacy while repurposing the land for community development. The demolition of above-ground infrastructure was completed in December 2019 (Figure 2). The demolition led to the removal of more than 100,000 tons of materials; 95% of the materials were recovered.

When Subiaco Oval in Perth was decommissioned, many of its features were sold at auction to preserve memories of its 50-year history (Figure 3). Some items, such as pieces of turf, goalposts, and seats, were highly sought after by sports fans and collectors, eager to own a part of Western Australia's sporting heritage. These auctions allowed locals to cherish memorabilia from the oval, which hosted countless historic football games and events.

After the demolition of Subiaco Oval, a large urban redevelopment project known as "Subi East" [27] has been underway in Perth. This project aims to transform 35 hectares of land into a vibrant mixed-use precinct. The old stadium was replaced with a range of developments, including residential, commercial spaces, and community amenities. Part of the site has been retained for public use, with the oval's playing field preserved for community sports and integrated into a new high school, Bob Hawke College. The school opened its first stage in 2020 and is designed to serve up to 2,000 students. Additionally, the heritage of Subiaco Oval, including the historic entry gates and elements of football history, has been incorporated into the new landscape . The broader Subi East precinct is set to include a mix of social and affordable housing, public art, and Noongar cultural trails, celebrating the local heritage while providing new housing options and public spaces.

5.1. Key Stakeholders

The case study has two major players including two public organisations. The following sections provide an insight into these organisations' operations including their sustainability drivers and their involvement in the Subiaco Oval project.

5.1.1. Development WA

This organisation is the State Government's central development agency, operating across WA with a diverse portfolio of industrial, commercial and residential projects. By working closely with government, partners, industry and the community, this organisation aims to provide a bridge between private industry and the WA Government to secure sustainable community outcomes. Central to this work is the identification, design and implementation of major land and infrastructure projects. This involves undertaking long-term, complex and challenging projects and collaborating with many different stakeholders to deliver the best outcomes for the State.

The organisational sustainable activities and initiatives are informed by four pillars of their Integrated Sustainable Development framework [28]. These pillars include community well-being, environmental responsibility, design excellence and economic health. This public organisation uses Global Reporting Initiative (GRI) Standards to reflect transparency, accountability and the continuous improvements of their activities. The GRI Standards represent global best practice for reporting publicly on a range of economic, environmental and social impacts.

5.1.2. Western Australia MainRoads

Main Roads WA has played a pivotal role in advancing the use of recycled materials in road construction and maintenance projects across the state. Their commitment to sustainability is reflected in various initiatives aimed at reducing the environmental footprint of infrastructure development. Main Roads WA has also collaborated with various stakeholders, including local governments and the building and construction sector, to create a robust framework for incorporating recycled materials into large-scale projects. Their work includes conducting trials and studies to ensure that the recycled materials meet quality and durability standards for road construction. The organisation has also supported training programs and knowledge-sharing initiatives that aim to build capacity within the industry for using recycled products. Through these efforts, Main Roads WA has become a leader in sustainable road-building practices, setting an example for other Australian states and contributing to the wider adoption of circular economy principles in public infrastructure projects.

This agency contributed to the development of the Roads to Reuse (RtR) program in WA to promote the use of PwRC. The program encourages the integration of high-quality PwRC, like crushed concrete, into road bases and embankments. Through stringent quality controls and testing, the program ensures that recycled materials meet the required safety and performance standards. This initiative supports both environmental sustainability and economic efficiency by reducing landfill waste and promoting resource recovery in the state's infrastructure projects.

Table 5. The regulatory documents that were referred to in this case project

Title	Main relevance	Used by
WA's Land Authority Act (1992)	It requires the case project stakeholders to take account of and balance social, economic and environmental outcomes	DevelopmentWA
Roads to Reuse Specifications	Specify the quality of PwRC to be used in the state road infrastructure projects	MainRoads WA, demolition contractors, recyclers and local governments
Metropolitan Redevelopment Authority Act 2011 [29]	By promoting sustainable redevelopment, the Act indirectly encourages better waste management practices within urban settings	DevelopmentWA
DevelopmentWA's Sustainable Development Framework	The sustainable waste recovery and the use of PwRC are aligned with the four pillars of this framework (i.e., community wellbeing, environmental responsibility, design excellence and economic health)	DevelopmentWA

5.2. Regulations, Policies and Guidelines

This section is concerned with the main regulations, policies and guidelines that were referred to by the project owners and other stakeholders involved in planning, execution and the use of PwRC emerging from this case project. Table 5 summarises these resources.

6. Case Study Findings

6.1. Interviewees Profiles

This section of the report is based on interviewing three key stakeholders in this project. As outlined in Table 6, the interviewees were heavily involved in planning and/or execution of the case project. C_2P_1 was involved in the planning phase of this project 15 years ago when the stadium was considered as a prospective project. The interviewee contributed to agency consulting activities to plan for the end of life of the stadium when the state government decided to build a new stadium in another location.

C_2P_2 joined the project 7 from the beginning of the execution phase as a development manager. The interviewee led the demolition of the stadium. C_2P_3 serves as a sustainability advisor at MainRoads Western Australia, focusing on waste management processes and advocating for the use of PwRC in road infrastructure projects.

Table 6. Summary of participants' profiles in the case study

Stakeholder type	Experience/expertise	Participant code
Manager, Strategy & Innovation	Worked for the agency for 30 years and has been involved in planning for Subiaco Oval project	C_2P_1
Project Delivery Manager	Extensive experience as a development and project manager; currently working in the DevelopmentWA's operations team	C_2P_2
Principal Advisor Sustainability Strategy	15 years of experience in the field of circular economy, with a strong focus on advocating for incentives to promote the use of PwRC	C_2P_3

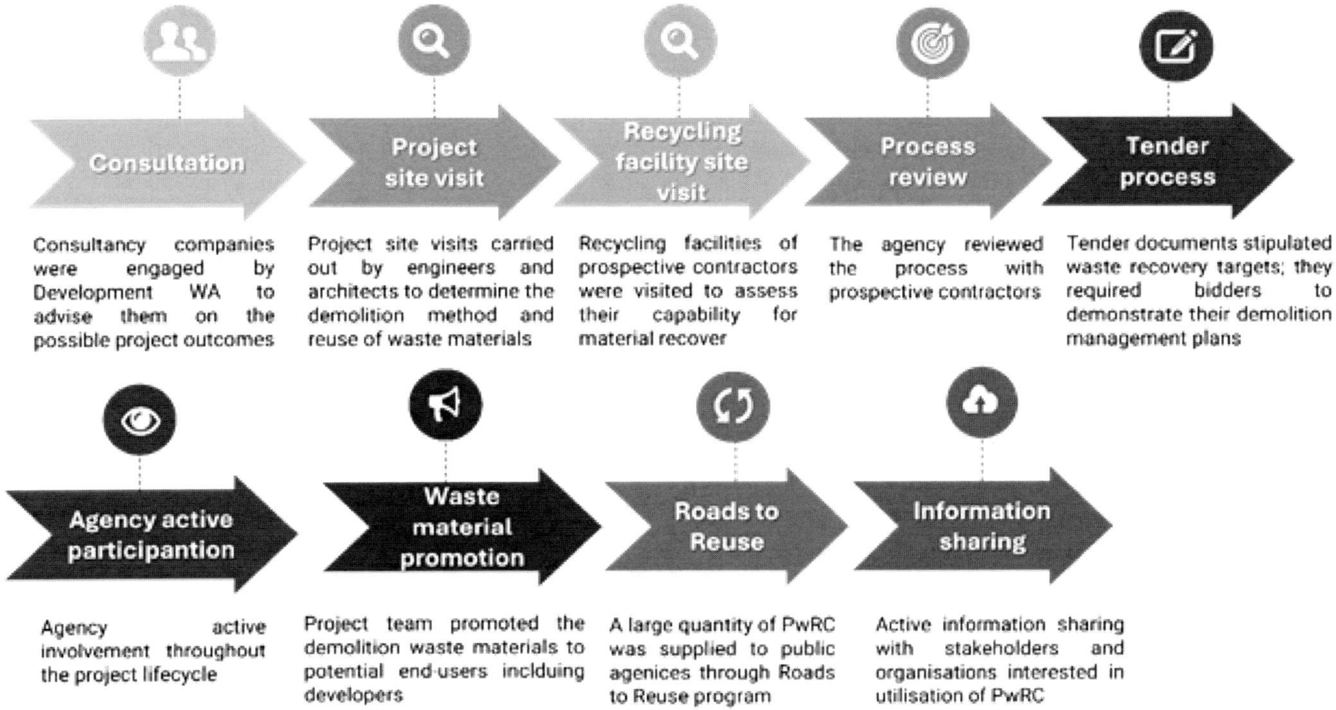

Source: Authors.

Figure 4. The interaction model for utilisation of PwRC in Subiaco Oval project.

6.2. The Interaction Model for the Utilisation of PwRC

This part of the report sheds light on the process of interaction between the key stakeholders to use demolition waste in construction projects. In the planning phase, the DevelopmentWA team engaged consultancy companies to advise them on the best possible outcomes of the project. The site visit to the project was carried out by an engineer and an architect to determine the best method of demolition of the stadium and the reuse of the resultant waste materials. Next, the agency team met with the prospective demolition and recycling contractors to go through the process of the project execution. This involved site visits to the project and the recycling facilities before awarding the contract. As a result, the waste recovery targets and requirements were included in the tender process which required the bidder to demonstrate it in their demolition management plans. The process is visualised in Figure 4.

7. Findings from Interviews

7.1. Key Challenges

The thematic analysis of the interviews identified 13 key challenges hindering the effective implementation of local collaboration for optimising the recovery of materials from the Stadium demolition waste and utilising the resulting PwRC in other construction projects. As presented in Table 7, these challenges were then categorised under three groups: project physical characteristics, project management, supply chain and policies and regulations.

7.1.1. Project Physical Characteristics

The physical characteristics of the demolition site appear to pose significant challenges to achieving the project's objectives. Specifically, these challenges have impeded both deconstruction efforts and on-site recycling activities. Additionally, interviewees highlighted two key issues regarding the handling of demolition waste materials: noise pollution and the presence of asbestos contaminants. The noise pollution was largely attributed to the site's location in a bustling suburb of Perth. Below are two illustrative statements from participants regarding the physical characteristics of the project:

Table 7. The key challenges identified in the interviews

Category	Challenge description
Project physical characteristics	• The old characteristics of the build made it difficult to deconstruct • The presence of contamination in demolition waste materials • Noise pollution for some on-site recycling activities • The physical characteristics of the project made it hard to reuse the extracted materials
Project management	• Cost implications of implementing additional management techniques • Difficulty in finding end users for demolition waste materials, particularly in local governments • Design warranty risk related to the use PwRC • Time constraints caused by the urgency to utilise the demolition site • The government's political pressure to deliver the project
Supply chain	• Limited experience within the construction team in utilising PwRC • Previous negative experiences with using PwRC in projects • Preference for working within the existing conventional supply chain • Uncertainty about the cost of the deconstruction process
Policies and regulations	• Unsupportive material specifications

'Because of the general shape of it being cyclic or however it is like being roundish, it was really hard to make good use of and like I said it was very specialised for what it was at the time' [C_2P_1].

'Through the project in terms of the management of the material, because there is the risk potentially for of contamination. So, therefore, you have to implement some different management techniques' [C_2P_3].

7.1.2. Project Management

Several challenges reported by the interviewees could be related to the project management aspects. These aspects included cost management, stakeholder management, risk management, and time management. In terms of cost management, the interviewees indicated that the higher costs for implementation of additional management techniques for handling materials present a challenge towards optimal use of PwRC through local collaboration. C_2P_3 expressed that *'Obviously there are potentially some extra costs or different costs through the project in terms of the management of the material because there is the risk potentially for of contamination. So, therefore, you have to implement some different management techniques for the product.'*

Moreover, participants' responses indicated that managing stakeholders for local collaboration proved to be challenging. This included addressing government requests for the project's delivery and identifying suitable end-users for the demolition waste materials. C_2P_2 reported that *'I was really trying to work with our local government to reuse on-site for the internal roads and at that time that wasn't successful.'* Another significant challenge was related to the project's tight delivery schedule. There was a pressing urgency to complete the project, as the land, once vacated, was earmarked for a major urban redevelopment into a mixed-use precinct. The need to align with the development timeline added considerable pressure to the project's execution. Some statements from participants highlighting the urgency of the work are provided below:

> 'We had to get this done and there was a very short time span of the obviously delivering the Oval so that it would time or align with the opening of the new college' [C_2P_1].

> 'There could be local government use because I know they had a section where they wanted part of that land for a community building' [C_2P_2].

The final challenge in managing the project was handling the risks associated with using PwRC in construction. Respondents pointed to uncertainties surrounding the performance of structures utilising PwRC. These materials require tailored design strategies to mitigate the risk of underperformance, making it difficult to guarantee both the design integrity and overall performance of the build.

7.1.3. Supply Chain

The supply chain challenges stemmed from the fact that, at the time of project execution, the use of PwRC had not yet become standard practice in the industry. As a result, there was uncertainty regarding the costs associated with deconstruction and how PwRC in new construction projects need to be utilised. C_2P_2 highlighted that '*The cost was a bit unknown, particularly with that sort of large scale.*' C_2P_3's response suggests the challenge of lack of experience with utilisation of PwRC: *'The construction team wasn't used to using it so that in itself brings its own challenges for the team itself and getting used to handling the product.' Previous negative experiences with the use of products with PwRC in other state projects likely contributed to the complexity.* C_2P_3 noted that '*I can't remember which project it was, but they had a bad experience with one of them and I think that they might have been a cross contamination...and it really did cause an issue for them, and they just hit the pause button for a while.'*

Additionally, the industry's preference for conventional supply chains, which had limited capacity to handle deconstruction waste materials and recycled products, further complicated the situation. As described by C_2P_3 '*there were commercial challenges, so supply chains of projects in many cases are already arranged to work together so therefore they'll seek to use the products that they have available to them and support their own supply chains.'*

7.1.4. Policies and Regulations

The participants noted that prior to the introduction of the Roads to Reuse materials specifications, local governments were unable to utilise these materials due to existing specifications that prohibited their use in local roads and other applications.

7.2. The Project's Key Sustainability Benefits

The completion of the project offered several sustainability benefits that are also aligned with several SDGs. These benefits were in the domain of environmental, social and economic sustainability. These benefits are visualised in Figure 5. According to the participants, achieving these benefits has opened the door for similar projects with comparable conditions and requirements across the state.

7.3. Drivers for Local Collaboration and PwRC Utilisation

The interviewees were also asked for their views regarding the main drivers for local collaboration to utilise PwRC in the case project. The respondents expressed eight drivers that are visualised in Figure 6. These drivers can be categorised into three groups: 'government sustainability commitment and demonstration,' 'sustainability recognition at the project level' and 'policies and regulations.'

As part of the state government, DevelopmentWA, the project owner, is legally obligated to align with the state's targets and ambitions to adopt innovative, sustainable and circular economy practices and incorporate PwRC in their construction projects. According to C_2P_1, '*It's an iconic sight and to some extent it's a demonstration project. So, it's got a lot of public attention and so innovation in whatever form we could and in whatever phases of the project we're taking, you know, being innovative in terms of how we deal with waste and recycling.*'

Source: Authors.

Figure 5. Sustainability benefits delivered in the case study.

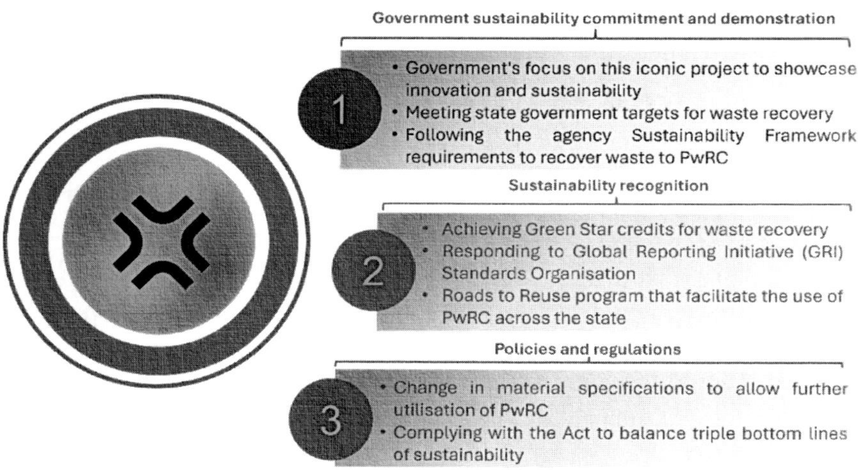

Source: Authors.

Figure 6. The main drivers supporting the local collaboration in the case project.

Table 8. The key strategies to address the challenges identified

Category	Strategy description
Effective project development and management	• Site visits with potential contractors • The tender process mandating demolition contractors to recycle waste generated from the project • Contractual requirement for using PwRC in state construction projects • Provision for risk-taking throughout the project lifecycle • Using technology to address dust pollution • Engagement of external environmental hazmat auditors to ensure PwRC quality • Effectively communicating with other agencies for adopting PwRC • Effective communications with prime contractors and subcontractors • Applying the learning from previous projects to avoid certain risks such as contamination • The agency (DevelopmentWA) active participation in the project • Reassuring the local community of the project's sustainability benefits
Organisational capacity enhancement	• Establishing clear alignment between all levels of the business • Running pilot trials prior to full implementation • Granting design concessions to facilitate the utilisation of PwRC in the state's construction projects • Providing training to the industry on the use of PwRC
Government support and policies	• Offering grants to support the use of PwRC in state's construction projects • Land Corp sustainability policies • Creating guidelines for local governments on the use of PwRC

Furthermore, achieving sustainability status on the project was a strong motivator, as it facilitated the use of demolition waste in other state construction projects. The project team was particularly driven to make this a reality, aiming to secure a Green Star rating for waste recovery, reflect these efforts in their GRI reports, and contribute to the state's Roads to Reuse program.

Lastly, the change in the material specifications allowing greater uptake of PwRC was a big driver, as it could convince the local governments to utilise the demolition waste in their local road projects. Also, by facilitating the use of PwRC and enhancing waste recovery, the project team were able to align their project objectives with the Metropolitan Redevelopment Authority Act

2011 [29] requirement to balance triple bottom lines of sustainability. One participant remarked '*Western Australian government has its own waste strategy, policy and implementation plan, and this project was specifically named and had its own set of recycling targets that we had to report on*' [C_2P_1].

7.4. Key Strategies to Enhance Local Collaboration in Using PwRC

There were several strategies that were employed to enhance local collaboration aiming to optimise the use of PwRC emerged from the demolition project. These strategies fall into three categories: 'effective project development and management,' 'organisational capacity enhancement' and 'Government support and policies.' Table 8 summarises these strategies.

7.4.1. Effective Project Development and Management

This category includes numerous strategies aimed at promoting locally collaborative efforts to boost the use of PwRC. These approaches have proven effective, particularly since many can be integrated during the project's planning stage, where objectives and management methodologies are established.

The three first strategies focus on the method of engaging head contractors in circular economy initiatives embedded in the state projects. These involve: (1) mandatory requirements for contractors to recycle demolition waste, (2) contributing to the facilitation of utilising PwRC in other construction projects, and (3) conducting site visits prior to bid acceptance or contract awards.

The next three strategies fall under risk management, where the project team responsible for developing the project management plan (PMP) should allow for an acceptable level of risk throughout the project lifecycle. The participants also indicated that, regarding environmental risks, utilising advanced technology to manage dust pollution (e.g., mist technology) and engaging external hazmat auditors can be beneficial. C_2P_2 noted that '*We undertake the auditing process as we go through to remove hazardous materials. So, we do have those additional steps to guarantee that we do not contaminate any of the recycling products.*'

Communication management was reported as a crucial factor for success in projects involving PwRC. According to the participants, effective communication with head contractors, subcontractors, and other state agencies is essential to ensure that the project's sustainability and circularity objectives are achieved. This communication management should also extend to include information and knowledge-sharing activities that aim to learn from and avoid certain risks encountered in previous projects.

The other two strategies relate to stakeholder management, which requires project owners to become active participants throughout the project lifecycle, interacting with various stakeholders, including local communities that could potentially be affected by project execution. According to C_2P_1 '*They [the community] would have understood that we're driving towards recycling. In fact, they probably would have been curious as to what we were doing with the material. If we had said, we're just going to bury them in a landfill, I think you would have seen them protesting about that.*'

7.4.2. Organisational Capacity Enhancement

Some key strategies identified in the interviews focus on enhancing organisational capacity to manage projects with circularity objectives. These strategies include establishing clear alignment across various sections of the organisation and conducting pilot trials of PwRC before large-scale application. Participants also emphasised that organisations should be building the capacity to provide education and training to the industry on the use of these resources, as well as grant design concessions, both of which facilitate the utilisation of these resources in alignment with the organisation's sustainability objectives. C_2P_3 noted that '*education is necessary for the recycling industry, to allow them to produce a product that was to a high standard and to give them an understanding of the scheme itself, scheme of the Roads to Reuse scheme.*'

7.4.3. Government Support and Policies

Key strategies under this category include offering monetary incentives for the use of PwRC, implementing sustainability policies set by Land Corp, and developing guidelines for local governments. These guidelines aim to empower local authorities to foster collaboration and increase the adoption of PwRC in their construction projects. C_2P_3 remarked '*We're also developing guidelines for the local governments to overcome the technical. Capability gap that might be there. To assist with that decision making and broader uptake.*'

7.5. Future of Local Collaboration for Application of PwRC

In response to the question regarding future trends for the optimal uptake of PwRC through local collaborative initiatives, participants identified seven contributing factors. These trends are summarised in Table 9.

As local governments have become more supportive, the use of PwRC is expected to increase in the areas where these materials are generated. C_2P_1 explained that *'I think also the general acceptance of recycled materials, standards have changed. Local governments have been more receptive, state governments are more receptive.'* Furthermore, the focus is shifting toward local utilisation of these resources to reduce emissions associated with truck movements from demolition and construction sites.

Table 9. Future trends of using PwRC through local collaboration

Trend	Description
The important role of local governments	As local governments become more supportive, the use of PwRC is expected to rise in areas where these materials are produced.
Focus on local utilisation	Efforts are shifting toward local utilisation of PwRC to reduce emissions from transporting materials from demolition and construction sites.
Iconic projects as catalysts	High-profile projects like Subiaco Oval are helping promote the local adoption of PwRC in building and construction
Information sharing enhances advocacy	Transparent information-sharing mechanisms, such as industry submissions and sustainability award pitches, are empowering key stakeholders to advocate for PwRC use.
Cost advantages	In the future, there will likely be cost advantages to using PwRC due to the proximity of recycling facilities to construction projects.
Broader range of PwRC	Local collaboration is expected to expand to include a wider range of PwRC, such as glass and plastic.

Through iconic projects like Subiaco Oval and transparent information-sharing mechanisms - such as industry submissions and pitching for sustainability awards - informed key stakeholders are now able to advocate for the local adoption of PwRC in building and construction projects. As highlighted by one participant *'communicating the nuts and bolts of how things have occurred, making submissions to industry or pitching projects for awards; that's all done deliberately around communication, external communications to stakeholders, shareholders and industry to try and change practise. It's done with clear intent'* [C_2P_1]

One participant also indicated that the in future there will be cost advantages for using PwRC due to the proximity of recycling facilities to construction projects. Lastly, local collaboration for using PwRC will expand to include a broader range of materials, such as recycled glass and plastic. In the words of one participant. *'Other types of material will come online. I think recycled plastic, crushed concrete, recycled glass are different materials. That could benefit from the adoption for having recyclers within the state or within the city at this point in time'* [C_2P_3].

7.6. Sustainability Demonstration Mechanisms

The project team and their partners utilised two key demonstration mechanisms to showcase the project's sustainability achievements. These included obtaining a 6-Star Green Star rating provided by the Green Building Council of Australia (GBCA), which represents world leadership in environmental and social sustainability practices. In addition, the project earned an Infrastructure Sustainability (IS) rating awarded by the Infrastructure Sustainability Council (ISC). The IS rating evaluates the sustainability performance of infrastructure projects across various categories, such as environmental management, resource use, and social outcomes, ensuring a holistic approach to sustainable infrastructure development.

Conclusion

The key takeaway from this case study is that local collaboration is poised to become the most significant trend in the future of using PwRC in construction projects. By fostering partnerships between local governments, industry stakeholders, and recycling facilities, communities are better positioned to utilise locally generated materials efficiently. This approach not only helps reduce transportation emissions but also encourages sustainable practices in construction, making local collaboration a cornerstone for the broader adoption of circular economy principles in the built environment.

However, the process of local collaboration for the purpose of increasing the use of PwRC in construction projects requires careful attention to detail and the implementation of strategies that mitigate potential risks. According to this case study, these risks span several domains, including environmental, financial, business, political, technical, and material and construction performance-related risks. Each of these areas must be addressed to ensure that collaboration efforts lead to sustainable and effective outcomes. For instance, environmental risks such as dust pollution or inadequate waste handling need to be managed alongside technical and performance-based concerns like the quality and durability of recycled materials used in construction.

The key strategies to mitigate risks associated with local collaboration in using PwRC can be categorised into three main areas:

1. Effective project development and management
2. Organisational capacity enhancement
3. Government support and policies.

These strategies work together to create a robust framework that promotes the sustainable use of PwRC in construction projects while mitigating various risk. Furthermore, the findings of this case study contribute to three main areas as indicated below. These contributions highlight the broader implications of the case study for advancing sustainability in the construction sector and fostering a circular economy.

Policy development- The insights gained can inform policymakers on the importance of supporting local collaboration initiatives and providing frameworks that encourage the use of recycled materials in construction. By understanding the challenges and successes highlighted in the study, policymakers can create more effective regulations and incentives that promote sustainability.

Industry practices- The case study's findings serve as a valuable resource for the building and construction sector stakeholders, offering practical strategies and best practices for integrating PwRC into projects. This can lead to enhanced awareness of the benefits and feasibility of using recycled materials, ultimately driving industry-wide change toward more sustainable construction methods.

Knowledge development- The research highlights the critical role of local collaboration in the effective uptake of PwRC within construction projects. By providing overall insight into the interactions among stakeholders, including project owners, state agencies, local governments, contractors, and communities, the study provides a nuanced understanding of how these relationships can influence sustainability outcomes. This adds to existing literature on collaborative governance and community engagement in sustainable development.

References

[1] DevelopmentWA. *The Subiaco Oval and Railway Precinct Design Guidelines*. 2021 [cited 2024; Available from: https://bit.ly/41h67WB.

[2] Shooshtarian, S., Maqsood, T., Wong, P. S., Caldera, S., Ryley, T., Zaman, A. and Ruiz, A., Circular economy in action: The application of products with recycled content in construction projects: A multiple case study approach. *Smart and Sustainable Built Environment*, 2024. 13(2): p. 370-394.

[3] Chembessi, C., Bourdin, S. and Torre, A., Towards a territorialisation of the circular economy: the proximity of stakeholders and resources matters. *Cambridge Journal of Regions, Economy and Society*, 2024: p. rsae007.

[4] DCCEEW, *Australia's Circular Economy Framework: Doubling our circularity rate*, E. Department of Climate Change, the Environment and Water, Editor. 2024, Australian Government: Canberra, Australia.

[5] Shooshtarian, S., Maqsood, T. and Wong, P. S., *Policy Intervention of Waste Management*, in *Waste Management in the Circular Economy*. 2024, Springer. p. 77-104.

[6] Shooshtarian, S., Caldera, S., Maqsood, T., Ryley, T. and Khalfan, M., An investigation into challenges and opportunities in the Australian construction and demolition waste management system. *Engineering, Construction and Architectural Management*, 2022. 29(10): p. 4313-4330.

[7] Ramos, M., Martinho, G., Vasconcelos, L. and Ferreira, F., Local scale dynamics to promote the sustainable management of construction and demolition waste. *Resources, Conservation & Recycling Advances*, 2023. 17: p. 200135.

[8] EPA NSW, *Protection of the Environment Operations (Waste) Regulation 2014*. 2014.

[9] Directive, E., Directive 2008/98/EC of the European Parliament and of the Council of 19 November 2008 on waste and repealing certain Directives. *Official Journal of the European Union L*, 2008. 312(3): p. 22.

[10] Halvorsen, E. O. and Andersson, H., Optimizing environmental and economic aspects of collaborative transportation and logistics related to infrastructure projects–A case study from Norway. *Waste Management*, 2023. 156: p. 159-167.

[11] Wu, H., Zuo, J., Yuan, H., Zillante, G. and Wang, J., Cross-regional mobility of construction and demolition waste in Australia: An exploratory study. *Resources, Conservation and Recycling*, 2020. 156: p. 104710.

[12] Akhtar, M. N., Bani-Hani, K. A., Malkawi, D. A. H. and Albatayneh, O., Suitability of sustainable sand for concrete manufacturing-A complete review of recycled and desert sand substitution. *Results in Engineering*, 2024: p. 102478.

[13] Aslam, M. S., Huang, B. and Cui, L., Review of construction and demolition waste management in China and USA. *Journal of Environmental Management*, 2020. 264: p. 110445.

[14] Almeida, F., Vieira, C. S., Carneiro, J. R. and Lopes, M. d. L., Drawing a Path towards Circular Construction: An Approach to Engage Stakeholders. *Sustainability*, 2022. 14(9): p. 5314.

[15] Essex, J. and Whelan, C., Increasing local reuse of building materials. *Proceedings of the Institution of Civil Engineers - Waste and Resource Management*, 2010. 163(4): p. 183-189.

[16] Wang, Y., Luo, S., Fan, J. and Zhen, L., The multidepot vehicle routing problem with intelligent recycling prices and transportation resource sharing. *Transportation Research Part E: Logistics and Transportation Review*, 2024. 185: p. 103503.

[17] Jian, S.-M., Wu, B. and Hu, N., Environmental impacts of three waste concrete recycling strategies for prefabricated components through comparative life cycle assessment. *Journal of Cleaner Production*, 2021. 328: p. 129463.

[18] Hiete, M., Stengel, J., Ludwig, J. and Schultmann, F., Matching construction and demolition waste supply to recycling demand: a regional management chain model. *Building Research & Information*, 2011. 39(4): p. 333-351.

[19] Shooshtarian, S., Caldera, S., Maqsood, T. and Ryley, T., Using recycled construction and demolition waste products: A review of stakeholders' perceptions, decisions, and motivations. *Recycling*, 2020. 5(4): p. 31.

[20] Treloar, G. J., Gupta, H., Love, P. E. and Nguyen, B., An analysis of factors influencing waste minimisation and use of recycled materials for the construction of residential buildings. *Management of Environmental Quality: An International Journal*, 14(1) : p. 134-145.

[21] Walker, E., Redmond, J. and Wang, C., Waste recycling: local methods for successful interaction with small business. *International Journal of Environment and Sustainable Development*, 2008. 7(4): p. 363-382.

[22] Yadav, M. and Sinha, S., Waste to wealth: Overview of waste and recycled materials in construction industry. *Materials Today: Proceedings*, 2022. 65: p. 2042-2052.

[23] Wu, H., Zuo, J., Yuan, H., Zillante, G. and Wang, J., Investigation of the social and economic impacts of cross-regional mobility of construction and demolition waste in Australia. *Resources, Conservation & Recycling*, 190(2023) : p. 106814.

[24] Macedo, P., et al., Learnings from Local Collaborative Transformations: Setting a Basis for a Sustainability Framework. *Sustainability*, 2020. 12(3): p. 795.

[25] Hafizah Hammad Ahmad, K., Nabila, A., Noorlailahusna Mohd, Y., Azyyati, A., Azlyn Ahmad, Z. and Shehu, E.-R., Pathways to sustainability: empowering indigenous communities through recycling education. *Information Management and Business Review*, 16(3S) : p. 1086-1093.

[26] Hocking Heritage+ Architecture. *Subiaco Oval auction*. 2019; Available from: https://hockingha.com.au/.

[27] NDY, *Sustainability Report: Lots 1 & 2, The Oval Subiaco East – Mixed-Use Development UEM Sunrise*. 2024, NDY Group: Perth, Australia.

[28] DevelopmentWA, *Attachment 3 – Sustainable development at DevelopmentWA*. 2024.

[29] State of Western Australia, *Metropolitan Redevelopment Authority Act 2011*, W. A. Government, Editor. 2024.

Chapter 4

An Analysis of ecologiQ: The Victorian State Change Platform

Salman Shooshtarian
Peter SP Wong
and Tayyab Maqsood
School of Property, Construction and Project Management, RMIT University, Melbourne, Australia

Abstract

Establishing an agency dedicated to centralising efforts to drive the circular economy and enhance the use of products with recycled content (PwRC) in the building and construction sector is critical for transitioning to a circular economy in this sector. Such an agency could serve as a hub for coordinating collaboration, research and development (R&D) activities, policy reforms and implementation of circular economy principles and practices. Hence, if set up properly, this change agency can drive innovation, collaboration, and sustainability across the sector. Therefore, this chapter presents the findings of qualitative research that aimed to understand how the existing change agencies operate to optimise the use of PwRC in construction projects. This chapter explores a few circular economy change agencies operating across Australia (i.e., Green Industries SA, Waste Authority WA, Sustainability Victoria) and overseas, to provide insights into their performance, and strategies adopted to increase the use of PwRC in construction projects. Secondly, the chapter analyses a case study using ecologiQ as a successful model to achieve sustainability in the major transport infrastructure construction sector in Victoria. The program, being delivered under the Victorian Infrastructure Delivery Authority, is particularly pushing the state's

In: Policy, Practice and Education of the Circular Economy in Built Environments
Editors: Salman Shooshtarian, Peter SP Wong and Tayyab Maqsood
ISBN: 979-8-89530-623-9

agenda for using PwRC in the major transport infrastructure sector. The program uses six major strategies to influence stakeholders to optimise the use of PwRC in construction projects: 'Recycled First Policy,' 'innovation showcase series,' 'knowledge hub,' 'speed networking exchange,' 'reforms in recycled materials specifications,' and 'greener infrastructure conference.' Lastly, the chapter outlines a comprehensive set of recommendations designed to significantly augment ecologiQ's capacity to drive the transition to a circular economy within Victoria's major transport infrastructure construction sector.

Keywords: products with recycled content, infrastructure, recycled first policy, circular economy, Victoria

1. Introduction

Establishing an agency dedicated to centralising efforts to drive the circular economy and enhance the use of products with recycled content (PwRC) in the building and construction sector is critical for advancing sustainability in this sector. Such an agency could serve as a focal point for coordinating research, development, and implementation of circular economy principles and practices. By bringing together stakeholders from across the sector, including manufacturers, builders, architects, and policymakers, the agency can facilitate collaboration and knowledge sharing, leading to more innovative and sustainable building practices. Additionally, a centralised agency can help set industry standards and guidelines for the use of PwRC, ensuring consistency and quality in construction projects.

Furthermore, an agency focused on the circular economy in the building and construction sector can help address specific challenges and barriers that currently limit the use of these resources. These challenges include issues related to supply chain logistics, quality control, and consumer perception. By providing support and resources to address these challenges, the agency can help increase the adoption of PwRC in construction projects. This, in turn, can lead to significant environmental benefits, such as reduced waste generation and lower carbon emissions. Overall, establishing such an agency can drive innovation, collaboration, and sustainability in the building and construction sector.

This chapter aims to analyse how ecologiQ, a program being delivered under the Victorian Infrastructure Delivery Authority, has progressed towards influencing stakeholders to enhance the use of PwRC in the state's major transport infrastructure projects. The next sections present the objectives, and the research approach, followed by the findings of the literature review, interviews with key stakeholders and recommendations.

2. Objective

This chapter aimed to explore how change platforms and initiatives can optimise the use of PwRC in construction projects by conducting a literature review and examining a case study in Victoria, Australia. The two key objectives include:

1. To investigate the strategies employed by change platforms in Australia and globally to promote the adoption of PwRC in the building and construction sector,
2. To explore how ecoloqiQ in Victoria enhances the use of PwRC in the state infrastructure projects.

3. Research Approach

3.1. Review of Literature

This section of the chapter presents the research approach employed. First, the research team conducted an extensive literature review to offer insight into leveraging change agencies or programs at the global level. This is followed by conducting a study analysis of some change agents in Australia. The cases include Waste Authority in Western Australia, Green Industries SA in South Australia, and Sustainability Victoria, in Victoria. The sources used in this step included journal articles and government and industry reports that are freely available online. Care was taken to only capture the literature that describes the operation of change agents to reflect the current trends and developments.

3.2. Case Study and Validation

The next phase of this research focused on a qualitative case study analysis of the ecologiQ program, which was established in 2019. This initiative plays a key role in advancing Victoria's circular economy by promoting the use of locally reused and recycled materials while minimising waste in transport infrastructure projects. A variety of secondary sources were analysed to gain a comprehensive understanding of the program's operational model and its impact across the state of Victoria. The case study was subsequently validated through a review by a representative from the Victorian Infrastructure Delivery Authority, ensuring the accuracy and relevance of the findings.

4. Literature Review

This section of the chapter presents the analytical findings from the literature review, which aimed to shed light on the role of change agents in promoting the adoption of PwRC in construction projects. The analysis is organised into two key subsections: International Perspectives (Section 3.2) and Australian Perspectives (Section 3.3), offering a comprehensive view of global and local practices.

4.1. Platform for Change: International Perspectives

4.1.1. The UK National Industrial Symbiosis Programme (NISP)

The UK National Industrial Symbiosis Programme (NISP) is a major initiative aimed at improving resource efficiency by promoting collaboration between industries. NISP is the first national industrial symbiosis (IS) programme, that was created by the UK Business Council for Sustainable Development and managed by the nonprofit International Synergies [1]. NISP has received recognition for its achievements from bodies including the UN, European Commission, and World Wildlife Fund (WWF).

Launched in 2005, NISP facilitates the exchange of waste, materials, energy, and water between businesses, effectively turning one company's waste into another's resource. This approach, known as IS, fosters a circular economy by reducing waste and minimising the use of raw materials. NISP participants ranged from micro, small, and medium enterprises (SMEs) to large multinational corporations, spanning every sector of industry. This inclusive approach enabled businesses of all sizes to collaborate and exchange resources, fostering innovation and promoting the circular economy across diverse industries. NISP has successfully diverted millions of tonnes of waste from landfills, helping industries repurpose materials and reduce environmental impacts.

One of the key factors behind the programme's exceptional success was its effective management and delivery. The initiative was driven by a nationally coordinated strategy, supported by a local delivery structure. The NISP teams' in-depth knowledge of regional economic and environmental priorities allowed the programme to have a significant positive influence across the UK [2]. This initiative has consistently played a pivotal role in inspiring local, national, and international governments to accelerate their efforts toward enhancing resource efficiency and driving the transition to a circular economy.

According to Castellet-Viciano, Hernández-Chover [3], other key elements that have contributed to the success of this IS model include government funding to support the programme, a wide and diverse network of organisations, dedicated professionals focused on maximising the value and extracting optimal benefits from their business interactions, and the existence of a system for managing and exchanging information and data. The initiative operation is based on five major themes as outlined in Figure 1.

International Synergies (ISL) has developed proven industrial symbiosis methodology, tools and systems that can be applied at any scale and in any economy based around the world's first National Industrial Symbiosis Programme.

ISL has developed and implemented the world's leading facilitated industrial symbiosis model (NISP®), based on demand-led engagement with businesses and other organisations. NISP offers a successful blueprint that can be adapted and replicated globally. Since 2007, ISL has exported the NISP model to over 20 countries at both national and regional levels through its capacity-building framework. Furthermore, the organisation has provided strategic consultancy and industrial symbiosis readiness training to more than 10 additional countries, supporting the global adoption of sustainable resource management practices. This international expansion has demonstrated the flexibility and effectiveness of the NISP model in promoting resource efficiency and reducing waste across various economic and environmental contexts.

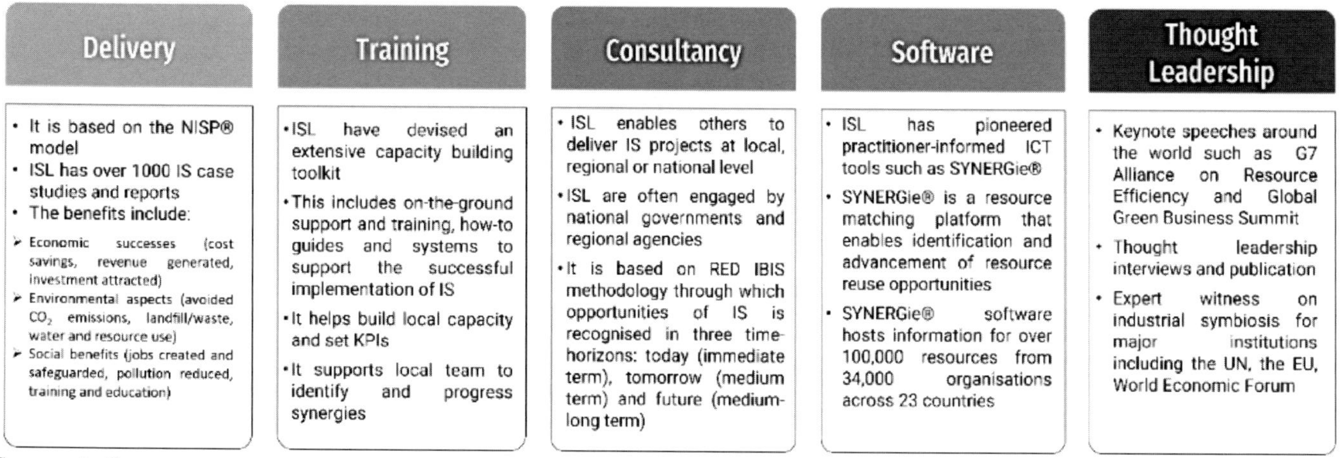

Source: Authors.

Figure 1. Major activities performed with the NISP initiative.

4.1.2. Platform for Accelerating the Circular Economy (PACE)

Platform for Accelerating the Circular Economy (PACE) is a global initiative aimed at promoting the circular economy by bringing together governments, businesses, and civil society organisations to collaborate and scale solutions that advance resource efficiency and sustainability [4]. PACE was created in 2018 by the World Economic Forum and is now hosted by the World Resources Institute. PACE is facilitated by a full-time team in The Hague, Netherlands.

Table 1. PACE's three activity pillars

Activity pillar	Description
Leadership	• PACE brings together a global leadership group dedicated to advancing the circular economy. • This group collaborates to address specific obstacles that hinder progress. • They pilot innovative projects and scale up successful practices to achieve a greater impact on sustainability efforts.
Learning	• PACE gathers and translates insights from its projects into frameworks and methods that can be replicated. • Knowledge is generated through mapping exercises across different thematic areas. • PACE facilitates cross-learning between projects to ensure lessons are shared widely. • These approaches are disseminated within the PACE community to guide leadership decisions. • The goal is to better enable the scaling of circular economy solutions by using these replicable models and shared learnings
Projects	• PACE initiates and accelerates new projects, while also helping partner-led initiatives scale circular economy efforts. • Focus areas include plastics, electronics and capital equipment, food and agriculture, and textiles and fashion. • Other thematic priorities involve Finance, Metrics, and Innovation to support the transition. • Active projects are being implemented across multiple regions, including Asia, Africa, Latin America, and Europe, showcasing the global reach of PACE's initiatives.

PACE is co-chaired by a group of global leaders, including government representatives and CEOs of multinational corporations such as Apple, SYSTEMIQ and IKEA. It also receives support from the World Economic Forum, which plays a key role in the platform's operations and outreach. PACE serves as a hub for sharing best practices, research, and data to help industries and governments adopt circular models. It provides toolkits, case studies, and policy recommendations to promote circular economy transitions globally.

The PACE current programs are informed by three priority areas to improve environmental quality, economic development, and social equity at the local, regional and country levels. These priority areas are illustrated in Figure 2.

Source: PACE (2024)[4].

Figure 2. PACE's three priority areas.

PACE works with governments to set ambitious circular economy targets and align policy with circular practices. For example, it supports countries in developing national strategies for circularity and sustainable development. In order to address the above-mentioned priority areas, PACE programs follow a process that involves four major stages These include background analysis, stakeholder analysis, running workshops and developing actionable recommendations (Figure 3).

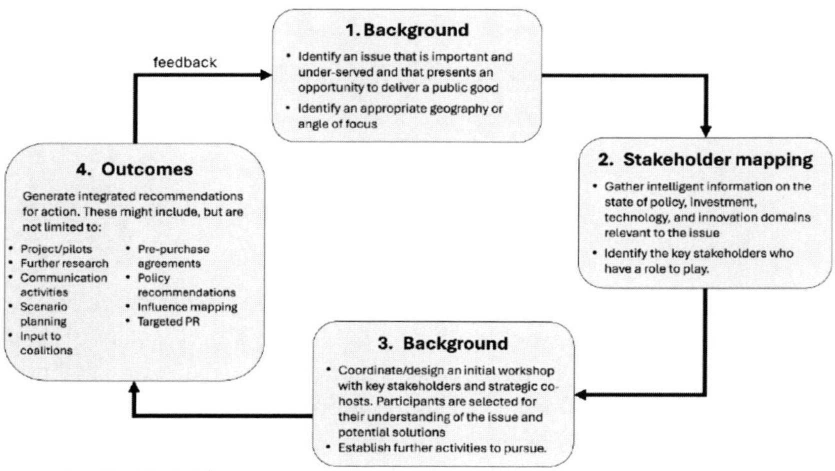

Source: PACE (2024)[4].

Figure 3. Schematic overview of the PACE process.

In alignment with Stage 4 (Outcomes), PACE has developed action agendas for specific industries, providing roadmaps to accelerate circular economy practices. These agendas focus on crucial sectors like electronics, plastics, and capital equipment. PACE's strategy focuses on three activity pillars: Leadership, Learning and Projects. The pillars closely interact with one another to implement and scale circular approaches.

4.2. A Platform for Positive Change: Australian Perspectives

In Australia, the waste is primarily regulated at the state level. The direction of change towards a circular economy and the use of PwRC in the building and construction sector is informed by the state Waste Strategy document [5]. In each state, the use of PwRC is promoted through several state agencies. However, in recent years, in line with increased awareness of waste management and the circular economy movements, some states have established centralised entities to transform waste management from a linear approach to a more circular one. These change agencies are typically the products of partnerships between different public organisations and in some cases private sectors. Figure 4 illustrates four change agents in four Australian states. In this section, the performance of there of these change agents is reviewed.

Source: Authors.

Figure 4. Australian states change agencies for optimal uptake of PwRC.

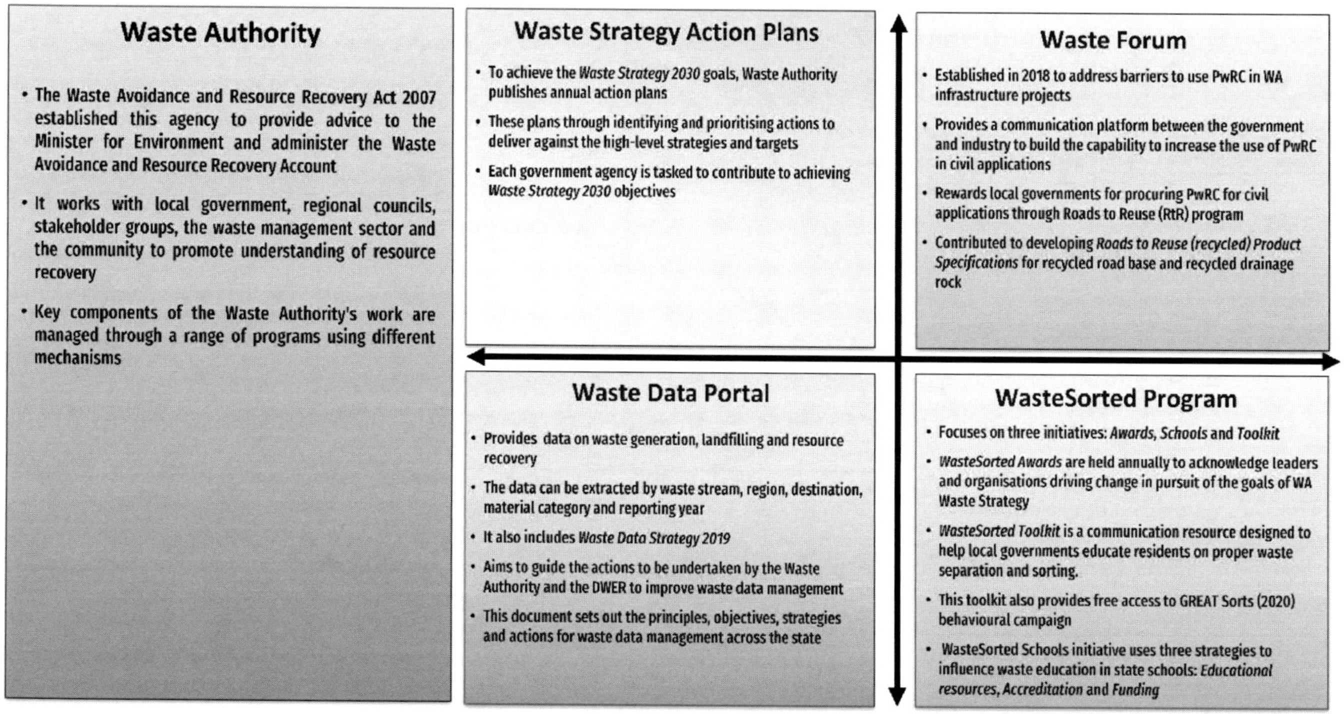

Source: Authors.

Figure 5. Waste Authority's major initiatives to enhance the use of PwRC in construction.

4.2.1. Waste Authority

Waste Authority is a government agency that works with local government, regional councils, stakeholder groups, the waste management sector and the community to promote understanding of resource recovery. The organisation uses four major initiatives to increase the use of PwRC in construction projects. As outlined in Figure 5, these initiatives include 'Waste Strategy Action Plans,' 'Waste Forum,' 'Waste Data Portal' and 'WasteSorted Program.

1. Waste Strategy Action Plans

In Western Australia, Waste Recovery and Resource Strategy is the overarching policy that drives the change in the C&D waste space. The state Department of Water and Environmental Regulation (DWER) is responsible for developing the policy in consultation with the Waste Authority and Main Roads Western Australia. The policy has been able to push the agenda of using PwRC at the state level. The policy is the foundation of the Waste Strategy 2030 document, which outlines the WA government mechanism for defining waste management-related actions.

Table 2. Headline strategies outlined in Action Plan 2022-23

Strategy	Description
1	Develop statewide communications to support consistent messaging on waste avoidance, resource recovery and appropriate waste disposal behaviours
2	A consistent three-bin kerbside collection system, which includes the separation of food organics and garden organics from other waste categories
3	Implement sustainable government procurement practices that encourage greater use of recycled products and support local market development.
4	Implement local government waste plans, which align local government waste planning processes with the waste strategy.
5	Review the scope and application of the waste levy to ensure it meets the objectives of the waste strategy and establish a schedule of future waste levy rates
6	Undertake a strategic review of WA's waste infrastructure (including landfills) to guide future infrastructure development.
7	Review and update data collection and reporting systems to allow waste generation, recovery and disposal performance to be assessed in a timely manner.
8	Provide funding to promote the recovery of more value and resources from waste with an emphasis on focus materials.

To achieve the Waste Strategy 2030 goals, Waste Authority publishes annual action plans. These plans through identifying and prioritising actions to deliver against the high-level strategies and targets. They are prepared with input from State Government agencies to support the Waste Strategy. According to the organisation's webpage [6], several action plans are used for this purpose. In Action Plan 2022-23, eight strategies are outlined to achieve Waste Strategy 2030 goals. These are summarised in Table 2.

In Waste Authority Business and Action Plan 2024-25, the most updated action plan developed by Waste Authority, each government agency is tasked to contribute to meeting Waste Strategy 2023. For instance, in Strategy No.3 (Table 2), the use of PwRC is requested from the Waste Authority, Department of Finance, Department of Communities, Department of Local Government, Sport and Cultural Industries, and Main Roads Western Australia.

2. Waste Forum

In Western Australia, the actions are defined in a platform called Waste Forum which was established in 2018. The Forum consists of *DWER*, *Waste Authority* and *MainRoads WA* representatives, as well as experts in the waste and recycling industry. Other parties involved include *Tyre Stewardship Australia*, the *Waste and Recycling Industry Association of Western Australia*, and the *State Department of Health and Finance*. The Waste Forum initiative has been the facilitator in addressing the long-standing impediments inhibiting the use of PwRC in WA infrastructure projects. This multi-disciplinary sustainability group has delivered outstanding outcomes through a commitment to innovation and collaboration. The platform allows the latter group to project manage the whole process to ensure that upcoming challenges as identified by them are dealt with appropriately.

One of the main policy-based actions for implementing sustainable procurement is to develop and improve recycled product specifications. In 2021, the forum contributed to developing Roads to Reuse Product Specifications that target recycled road-based and recycled drainage rock [7].

The Waste Forum also provides a communication platform between the government and industry to build the capability to decide on strategies to evaluate product specifications for PwRC. The platform brings together recycling businesses to promote and make the product specifications known to all parties involved in the recycled cornet supply chain. The Forum, in collaboration with the industry in the network, periodically facilitates training courses to educate the stakeholders about the requirements of PwRC. Such networking has enabled government organisations to connect to other waste groups, such as the Infrastructure Sustainability Council (ISC) WA Working Group, to capture the issues from a holistic perspective better.

Lastly, this platform worked with the state agencies to develop an incentive program called Roads to Reuse (RtR). The scheme provides a payment of $5 for each tonne of RtR product sourced from an accredited RtR supplier (Figure 6). The program budget for 2022–23 was $350,000. Payments will be made to local governments on a 'first come, first served' basis until the budget is exhausted.

To date, Main Roads has used over 100,000 tonnes of PwRC in projects [9] such as the Kwinana Freeway widening, Leach Highway and Welshpool Road Interchange, and Tonkin Gap, and is committed to significantly increasing their utilisation in future. This program also provides a list of accredited PwRC suppliers (recyclers) across the state. Currently, the list consists of four recycling companies.

3. Waste Data Portal

This online portal provides data on waste generation, landfilling and resource recovery. The portal through its interactive data dashboards allows one to select a combination of waste and recycling data that is of interest. The filters that can be used to navigate waste data include waste stream, region, destination, material category and reporting year.

This portal also includes the Waste Data Strategy 2019 [10] which aims to guide the actions to be undertaken by the Waste Authority and the DWER to improve waste data management in WA and to ensure that the waste data needs of stakeholders are met. This document sets out the principles, objectives, strategies and actions for waste data management across the state. The document is reviewed about every five years, in line with the legislated review of the Waste Strategy.

4. Waste Sorted Programs

This program focuses on three initiatives: *Awards, Schools* and *Toolkit*. The WasteSorted Awards are held annually and acknowledge exceptional leaders and organisations driving change in pursuit of the goals of Western Australia's Waste Strategy. The awards celebrate those making landfills the last resort by avoiding, reusing, repairing and recycling waste and inspiring others to follow in their footsteps. In 2024, the award program [11] consisted of ten categories to target individuals, business and industry, government, schools, community organisations and not-for-profits.

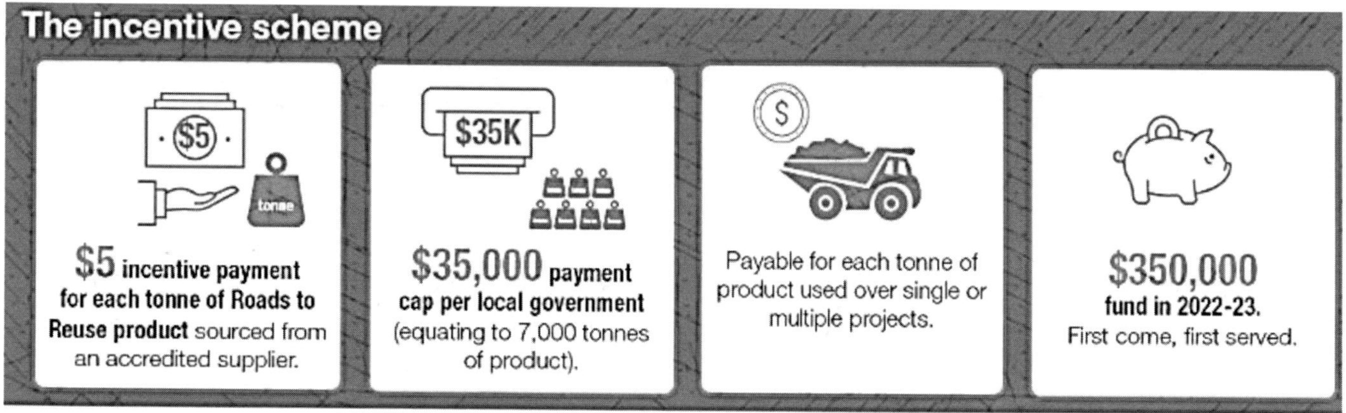

Source: Waste Authority [8].

Figure 6. Outline of RtR incentives mechanism.

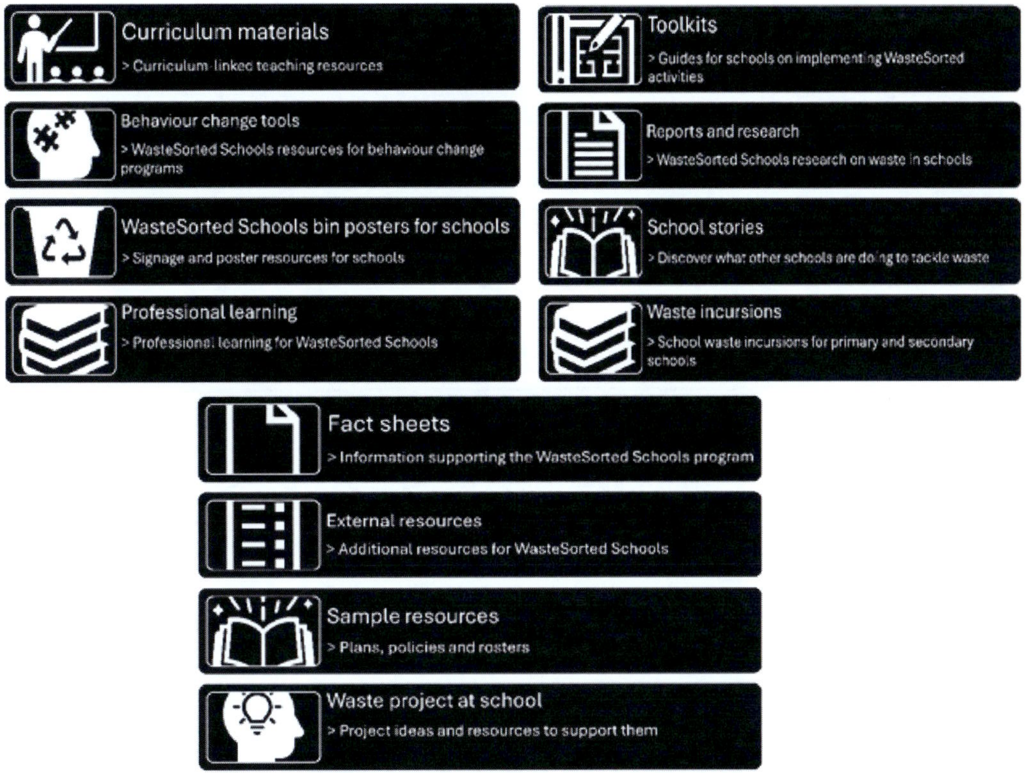

Source: Waste Authority [13].

Figure 7. The educational resources are provided through the WasteSorted WA Schools program.

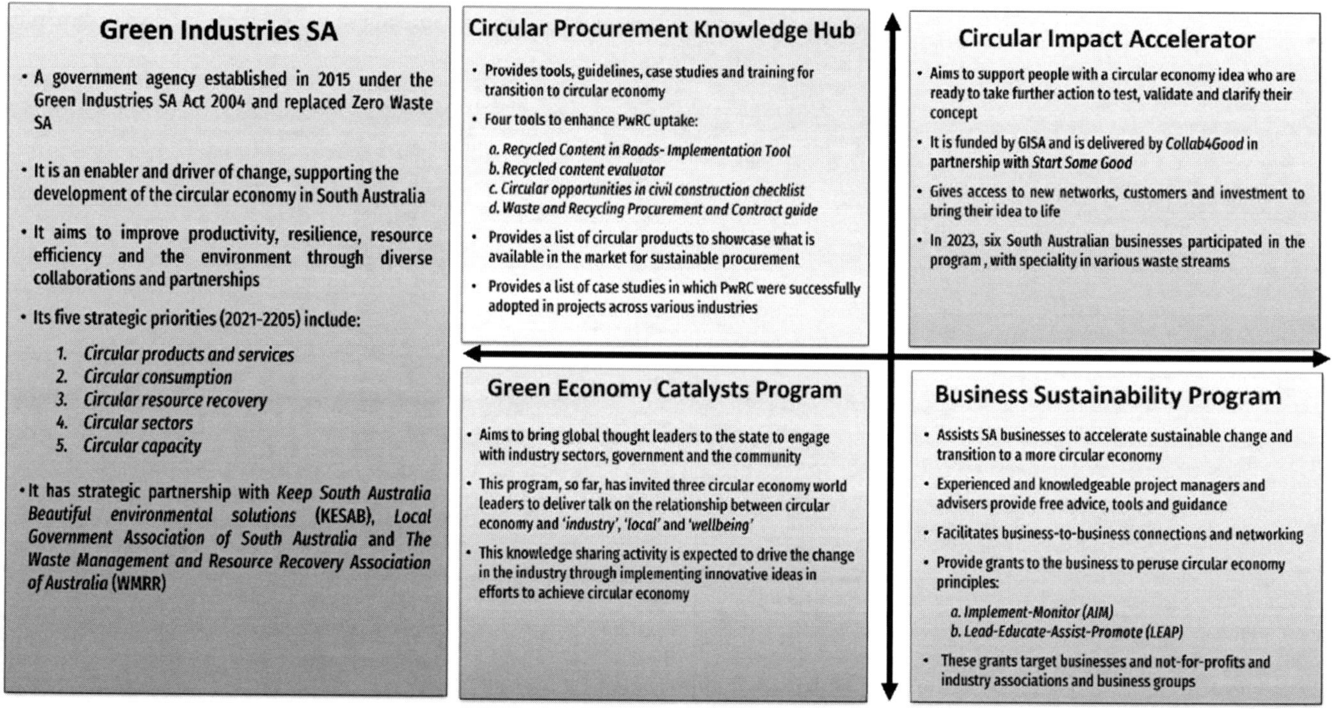

Source: Authors.

Figure 8. GISA initiatives in circular economy space.

WasteSorted toolkit is a communication resource developed to help local governments communicate with their residents on the importance of separating their waste and sorting it correctly to reduce contamination in their kerbside bin services. This source provides free access to a range of promotional content [12] such as templates and graphic elements, social media and videos, banners and displays for community events. Furthermore, an effective initiative under this program is the *GREAT Sorts* behavioural campaign. It was launched in 2020 and uses behavioural economics research and strategies developed locally for WA. The campaign development incorporates extensive consumer research and message testing to identify and prioritise waste sorting behaviours with the most potential to reduce household waste and increase recovery rates.

The WasteSorted Schools [13] initiative uses three strategies to influence waste education in state schools. It provides waste educational resources for WA schools, accredits schools with a waste education focus and provides funding to assist accredited schools in waste management and education. WasteSorted Schools has a wide variety of resources to help schools implement waste minimisation projects. The list of these resources is outlined in Figure 7.

4.2.2. Green Industries SA

Green Industries SA (GISA) is an enabler and driver of change, supporting the development of the circular economy in South Australia. This government agency was established in 2015 under the Green Industries SA Act 2004 [14] and replaced Zero Waste SA. This change agent strives to improve productivity, resilience, resource efficiency and the environment through diverse collaborations and partnerships [15]. GISA has outlined its strategic priorities through the GISA Strategic Plan 2021-2025 [16]. This strategic plan outlines how South Australia can ensure a sustainable future while maintaining a thriving economy. Green Industries SA's five strategic priorities will focus on Circular products and services, Circular consumption, Circular resource recovery, Circular sectors and Circular capacity over the next five years. This agency offers a range of initiatives that help the state transition to a circular economy and achieve its five strategic priorities (Figure 8).

1. Circular Procurement Knowledge Hub

One of the relevant initiatives led by this organisation is the Circular Procurement Knowledge Hub. This Hub provides tools, guidelines, case studies and training that will help incorporate circular economy principles into procurement and achieve sustainability goals.

This hub provides four tools that together assist businesses operating across the state in the effective utilisation of PwRC. The first tool is the *Recycled Content in Roads- Implementation Tool*. This implementation tool has been developed to provide a clear pathway for the use of recycled material in infrastructure, in an environmentally safe and responsible way. It is aimed at all levels of local government, including elected members, executives, procurement teams, sustainability officers, asset managers and engineers to help them understand the PwRC options available for inclusion in local roads, and allow them to specify their use with greater confidence. This tool- which was supported by GISA- was developed by the Australian Road Research Board in consultation with the Institute of Public Works Engineering Australasia, the City of Mitcham, the City of Burnside, the Port Pirie Regional Council, and the Adelaide Hills Council. The content of this tool is provided in Table 3.

The second tool is the *Recycled Content Evaluator* which is designed to assist with the evaluation of quotes and proposals during a procurement process, making it easier to include PwRC as part of the evaluation criteria. Responses to each of the questions in the evaluator should be requested from potential suppliers as part of the tender process. The results of this tool can be extracted for inclusion in procurement documentation. This tool was adapted from the Local Government Area (LGA) of SA Buying it Back project outputs.

The third tool is the *Circular Opportunities in Civil Construction Checklist* [18], which was developed by the Department for Infrastructure and Transport's Regency to Pym Street project, and provides a checklist to easily identify circular opportunities in civil construction projects.

Table 3. Structure of Recycled Content in Roads - Implementation Tool

Content	Description
Recycled material information	Common PwRC with proven performance in road networks, and their relevant applications and specifications
Life cycle assessments	Results from LCAs on innovative pavement designs incorporating PwRC
Environmental pathways	Environmental regulatory pathways, potential barriers to use, and compliance and mitigation measures for the use of PwRC
Implementation framework [17]	Key steps to design and execute field trials for incorporating PwRC in road construction

The fourth tool is the *Waste and Recycling Procurement and Contract Guide* [19]. It provides information on how and when to consider changes to waste and recycling collection services and contracts, and guidance on preparing tender documents that encourage positive financial, contractual and environmental outcomes.

This hub also provides a list of products that are deemed to be circular. This list aims to showcase what is available in the market for sustainable procurement. For instance, in the construction tab, eight products materials and products are introduced. These include aggregates, bitumen, bridges, cement and concrete, pavements and footpaths, walls, kerbs and edging, water and drainage. GISA invites businesses operating in circular manufacturing and supplying across the state to add their circular products to the list.

Lastly, GISA showcases a list of case studies with sustainability achievements. Some of these case studies successfully utilised PwRC in their projects. These case studies cover a range of industries including construction, textile, food and agriculture.

2. Circular Impact Accelerator

This initiative is funded by GISA and is delivered by Collab4Good in partnership with Start Some Good. It aims to support people with a circular economy idea who are ready to take further action to test, validate and clarify their concept, while also gaining access to new networks, customers and investment to bring their idea to life. In 2023, six South Australian businesses participated in the program [20], with a specialty in various waste streams.

3. Green Economy Catalyst Program

This program aims to bring global thought leaders to the state to engage with industry sectors, government and the community, to support SA's transition to a circular economy and take advantage of the economic growth opportunities. Thus far, this program has invited three circular economy world leaders to deliver a talk on the relationship between circular economy and industry, local and well-being. The knowledge shared by these experts is expected to drive change in the industry through implementing innovative ideas to increase the impacts and achieve sustainability goals while transitioning to a circular economy.

4. Business Sustainability Program

This program assists South Australian businesses and industry sectors to accelerate sustainable change and transition to a more circular economy [21].

By identifying and prioritising better practice materials and resource efficiency, waste management and resource recovery, and implementing circular economy principles, participants in this program can improve profitability, productivity, and environmental performance. Through this program, experienced and knowledgeable project managers and advisers provide free advice, tools and guidance. It also facilitates business-to-business connections and networking. Part of this program is to provide grants to businesses to peruse circular economy principles. Currently, two categories of grants target businesses and not-for-profits (*Assess-Implement-Monitor (AIM) Grants),* and industry associations and business groups (*Lead-Educate-Assist-Promote (LEAP) Grants).*

4.2.3. Sustainability Victoria

Established in 2005, under the Sustainability Victoria Act 2005, Sustainability Victoria is a statutory authority with a board appointed by the Minister for Energy, Environment, and Climate Change Action. Its primary objective is to drive Victoria's transition to a circular, climate-resilient, and clean economy [22], aligning with the government's targets for 2025 and 2030 as outlined in the Recycling Victoria policy and the Climate Change Strategy. According to *the Sustainability Victoria Strategic Plan 2024-207* [22], this agency is a circular economy transition broker with the intent of reducing or stopping waste before it starts, in every part of the system. The agency's vision is to position Victoria as the leading circular economy state in Australia by 2027.

The agency's latest document, SV Strategic Plan 2024-2027 [22], outlines the major priorities towards achieving a circular economy. These include 'Investment & Innovation,' 'Behaviour Change & Education,' and 'Community Action.' Sustainability Victoria collaborates with a wide array of stakeholders, including industry partners businesses, entrepreneurs, research institutions, schools, households, individuals, community groups, and government agencies within Victoria and across Australia. By fostering these partnerships, Sustainability Victoria aims to develop and deliver impactful solutions that are tailored to meet the specific needs of each sector, contributing to the broader goals of sustainability. Sustainability Victoria closely works with ecologiQ to enhance the uptake of PwRC in the state's major transport infrastructure construction sector. This agency uses various strategies and initiatives to support the state's transition to a circular economy. Figure 9 presents these strategies:

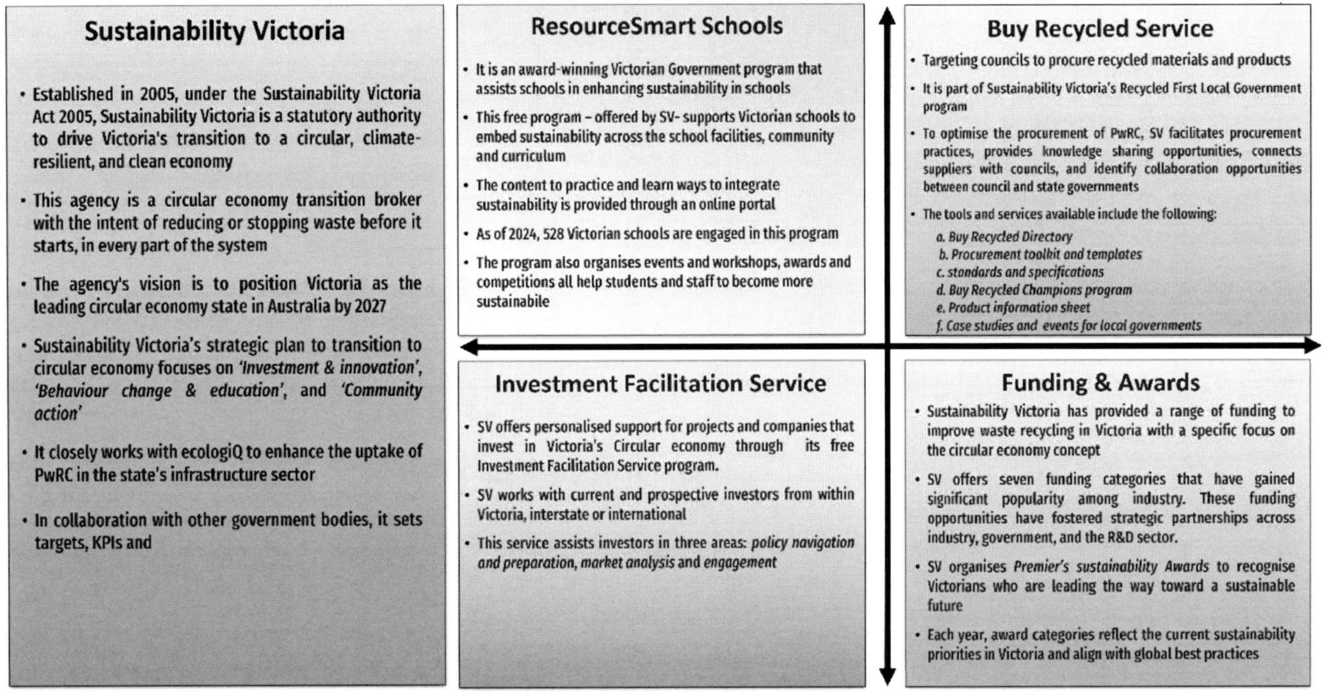

Source: Authors.

Figure 9. Sustainability victoria initiatives in circular economy space.

1. ResourceSmart Schools

ResourceSmart Schools is an award-winning Victorian Government program that assists schools in embedding sustainability in everything they do. This free program is offered by Sustainability Victoria and supports Victorian schools to embed sustainability across the school facilities, community and curriculum while saving resources and money for the school [23]. Schools join the program and then have access to an online portal (ResourceSmart Schools online), where they learn and practice ways to integrate sustainability. SV has engaged hundreds of Victorian schools in sustainability through ResourceSmart Schools, with 528 currently engaged [22]. The program also organises events and workshops, awards and competitions. Through their 'waste module,' schools learn how to implement operational practices focused on reducing, reusing, and recycling to minimise waste sent to landfills.

Table 4. Tools and resources available through Buy Recycled Service

Tool/resource	Aim
Buy Recycled Directory	lists PwRC to provide buyers with easy access to suppliers and product options.
Procurement toolkit and templates	The Buy Recycled Procurement Toolkit [24] and templates provide practical guidance to council officers seeking to procure recycled products and materials for infrastructure, landscaping, parks and gardens.
Standards and specifications	Standards and specifications outlining the characteristics of PwRC required for application in civil applications
Case studies and news	Content about councils who have used PwRC in their projects
Product information sheet	An overview of PwRC for councils considering their use in infrastructure, landscaping, parks and gardens.
Events for local governments	Provides learning from local government project managers who are enabling the circular economy through buying recycled and sustainable procurement.
Buy Recycled Champions program	Through this initiative, Champions are empowered to embed sustainable procurement practices as business as usual at their councils and provide participants with opportunities to build capacity, influence, network and share successes within their own councils and beyond.

2. Buy Recycled Service

The Buy Recycled Service and dedicated team work with councils to encourage more use of PwRC in infrastructure, landscaping, parks and gardens. This service is part of Sustainability Victoria's Recycled First Local Government program. Under this program, SV provides several services including facilitating procurement practices, knowledge sharing, connecting suppliers with councils, and identification of collaboration between council and state governments to optimise the procurement of PwRC. Table 4 presents the tools and resources available through this program.

3. Investment Facilitation Services

Sustainability Victoria offers personalised support for projects and companies that invest in Victoria's Circular economy through its free Investment Facilitation Service program. SV works with current and prospective investors from within Victoria, interstate or international [25]. To provide investors with a comprehensive level of service, SV coordinates across government including investment specialists within the Department of Jobs, Skills, Industry and Regions (DJSIR) and Invest Victoria for overseas companies looking to set up operations in Victoria for the first time. This service assists investors in three main areas: *Policy navigation and preparation*, *Market analysis* and *Engagement* (Table 5).

Table 5. The structure of SV's investment facilitation service program

Policy navigation and preparation	Market analysis	Engagement
Help you navigate the government's policy direction and its priorities	Provide information and market insight on waste and resource streams	Advise on how to engage with local communities
Inform you of environmental requirements and potential risks	Provide insight on the status and development of local end markets	Connect the investors with research institutions and potential partners
Offer business case development support		Identify potential financial and non-financial support
Identify suitable sites for your project and navigate approval processes		

4. Funding and awards

Since 2021, Sustainability Victoria has provided a range of funding to improve waste recycling in Victoria with a specific focus on the circular economy concept. So far, the agency has offered seven funding categories that have gained significant popularity among industry, government, and research institutions (Table 6). These funding opportunities have fostered strategic partnerships across industry, government, and the R&D sector.

Furthermore, Sustainability Victoria organises the annual Premier's Sustainability Awards to recognise Victorians who are leading the way toward a sustainable future. Each year, the award categories reflect the current sustainability priorities in Victoria and align with global best practices. The latest awards round in 2024 includes categories such as Circular Economy Innovation.

Table 6. The categories of SV's funding and awards

Premier's Sustainability Awards categories	Funding and grant categories
1. Circular economy innovation	1. Innovation
2. Future energy	2. Business support
3. A healthy and fair society	3. Communities
4. Sustainable places	4. Research and development: materials
5. Thriving environment	5. Waste to Energy
6. Waste and recycling solutions	6. Household education
	7. Recycling modernisation fund

4.2.4. Office of Environment and Heritage NSW

The Office of Environment and Heritage (OEH) was formed in 2011 and is now part of the NSW Department of Climate Change, Energy, the Environment and Water. This public agency worked with communities, businesses and governments to protect, preserve and strengthen the quality of NSW's natural environment and heritage. OEH is responsible for delivering the state outcome 'connecting communities to resilient and sustainable environments and heritage.

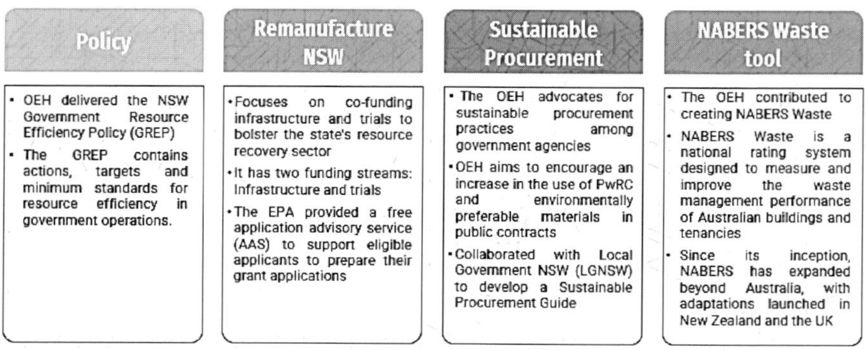

Source: Authors.

Figure 10. The main OEH's activities concerning the use of PwRC.

The OEH in NSW is actively involved in promoting the use of recycled materials through various initiatives, policies and programs aimed at enhancing resource recovery and supporting sustainable practices. These activities are captured in Figure 10.

1. Policy development

OEH delivered the NSW Government Resource Efficiency Policy (GREP) through the former Cluster Corporate Services (CCS) Sustainability Program. The GREP contains actions, targets and minimum standards for resource efficiency in NSW government operations.

2. Remanufacture NSW

One of the key programs is Remanufacture NSW [26], which focuses on co-funding infrastructure and trials to bolster the state's resource recovery sector. This program was developed in response to Australia's waste export bans and aims to increase recycling and reuse activities while keeping materials within the economy. Remanufacture NSW is jointly funded by the Australian Government's Recycling Modernisation Fund and the NSW Government's Waste Less, Recycle More initiative. The grant program is delivered by the NSW Environmental Trust in partnership with the NSW Environmental Protection Authority (EPA).

The initiative specifically targets materials like plastics, paper, cardboard, and tyres, supporting projects that enhance recycling infrastructure and develop new processing technologies. Remanufacture NSW had 2 streams: infrastructure and trials. Under Stream 1 Grants are available for establishing new recycling facilities, improving sorting and cleaning technologies, and increasing processing capacities for various recyclable materials. Under Stream 2, Funding is also allocated for innovative trials that aim to improve the use of PwRC in manufacturing and construction. This includes testing new methodologies, technologies, and processes to increase the quality and marketability of PwRC.

3. Sustainable procurement

The OEH advocates for sustainable procurement practices among government agencies to increase the use of recycled and environmentally preferable materials in public contracts. One significant effort in this space is the collaboration with Local Government NSW (LGNSW) to develop a Sustainable Procurement Guide [27]. This guide helps councils maximize their purchasing power - over $10 billion annually - by encouraging them to adopt sustainable practices that benefit both the environment and the economy. This partnership has focused on training, capacity building, and developing best practices to embed sustainable procurement into the daily operations of local councils.

Additionally, OEH was instrumental in the launch of the Sustainable Procurement ISO 20400, the first international standard for sustainable procurement. This standard serves as a framework to help organizations implement sustainable procurement strategies effectively.

4. NABERS Waste

The National Australian Built Environment Rating System (NABERS) was developed in NSW, starting in 1999. Its creation was driven by the NSW government, specifically through the OEH, in collaboration with various stakeholders, including the property industry and sustainability experts. Since its inception, NABERS has expanded beyond Australia, with adaptations launched in New Zealand and the UK, demonstrating its influence in promoting sustainable practices globally. The program encourages building owners and tenants to enhance their operational efficiencies and minimize their environmental impact by providing transparent performance ratings.

NABERS Waste is part of a broader initiative to encourage and recognise sustainable practices in building management. It focuses on assessing the volume of waste generated, the effectiveness of recycling practices, and the overall waste management strategies employed by properties. The system not only provides a rating but also guides building managers and operators on how to improve their waste management practices. This can include implementing better recycling systems or reducing overall waste generation.

5. Case Study: ecologiQ

5.1. Background

The ecologiQ program was initiated in 2019 by Allen Garner, the CEO of Major Road Projects Victoria. Garner recognised the critical link between Victoria's transport projects and the need for reform in the state's waste recovery system. His pioneering vision laid the foundation for ecologiQ's efforts to drive sustainability and innovation in Victoria's major transport infrastructure projects. The four main objectives of this program are as follows [28]:

- Contribute to a Victorian circular economy by increasing the use of Victorian reused and recycled materials, and reducing waste in transport infrastructure construction
- Encourage innovation in transport infrastructure construction to improve quality and accelerate the implementation of new Victorian recycled products
- Support a vibrant and sustainable Victorian market for reused and recycled materials in transport infrastructure construction
- Reduce, re-use and recycle materials without compromising the delivery of safe, high-quality and sustainable infrastructure.

A significant catalyst for this transformation is the Recycled First Policy, a groundbreaking initiative in Australian history. This policy mandates that contractors involved in Victorian transport projects optimise the use of recycled and reused materials, marking a crucial step towards sustainable construction practices in the region. Since the policy was implemented in 2020, more than 4.5 million tonnes of recycled materials have been delivered across 46 projects – more than enough to fill the Melbourne Cricket Ground. In 2023, the ecologiQ program was the winner of Victoria's Premiers' Sustainability Award in the category of Circular Economy Innovation [29] for its efforts to drive greater use of PwRC in transport infrastructure projects.

5.2. Change Strategies

ecoloqiQ has implemented a variety of strategies to promote the adoption of PwRC in state major transport infrastructure construction projects. These strategies have been developed in collaboration with key state agencies such as Sustainability Victoria, Recycling Victoria, the Department of Transport and Planning, and Victoria's Big Build project offices, as well as various industries involved in C&D waste management. Figure 11 provides a visual representation of these strategies, highlighting the comprehensive approach taken by ecoloqiQ to drive positive change in the industry.

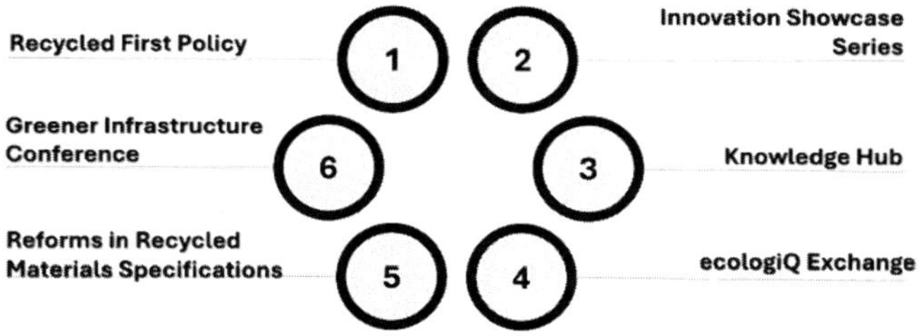

Source: Authors.

Figure 11. ecologiQ change strategies to optimise the use of PwRC in major transport infrastructure projects.

Recycled First Policy [30]-The Recycled First Policy aligns with the Victorian Government's Circular Economy Strategy, Recycling Victoria: A New Economy [31], a 10-year plan to revamp the state's recycling sector, enhance domestic recycling capabilities, and drive innovation. Under this policy, since March 2020, all bidders for major transport projects in Victoria must demonstrate in their proposals through a Recycled First Plan how they will optimise the use of PwRC within the constraints of existing standards and specifications to build roads, railways and other major transport infrastructure. The delivery agencies mainly include the Victorian Infrastructure Delivery Authority (VIDA) and the Suburban Rail Loop Authority (SRLA). ecologiQ facilitates the implementation of the Recycled First Policy, and thus PwRC, into Victoria's major transport infrastructure construction projects, including collating and analysing data from projects to identify PwRC use, barriers and learnings, then acting on and communicating these insights to delivery partners. The policy also has the potential to be extended to other projects, operational activities, and maintenance tasks undertaken by various State Government departments and agencies.

Innovation Showcase Series – The ecologiQ Innovation Showcase series gives Victorian industry and their research partners the chance to pitch ideas and products that can help transition Victoria towards a waste-free, low-carbon, climate-resilient future (Figure 12).

Moving towards a circular economy involves finding new uses for end-of-life materials and solutions to reuse, repurpose and repair products to divert waste from landfills. The team seeks pitches from both industry and research partners, using Victoria's Big Build as a catalyst for new markets. This presents an opportunity to showcase a new product or innovation relevant to major road and rail projects and escalate ideas to a captive audience. The pitches are reviewed by an expert panel comprising senior Victorian rail, road, and transport department representatives.

Knowledge Hub – The ecologiQ Knowledge Hub is an online site made available to Contractors, Suppliers, Project Teams and Stakeholders, which provides a central location to access a range of resources to assist with implementing PwRC. Available resources include an interactive PwRC Supplier Map; an Innovation Register, showcasing innovative PwRC research, trial and implementation status; Recycled First Policy documents, including templates, agency procurement guidance and reporting user guides; Visual Guides, showing opportunities to use PwRC in transport infrastructure; Reference Guides, summarising standards and specifications relevant to implementing PwRC; training videos and recorded sessions; a forecast PwRC Demand Model; and an overview of funding and assistance opportunities.

ecologiQ Exchange – The ecologiQ Exchange is an online networking event that helps drive innovation and collaboration across the industry, with ecologiQ's buyer/supplier exchanges helping connect contractors, advisors and project teams to the recycled material supply chain and giving suppliers an understanding of project needs.

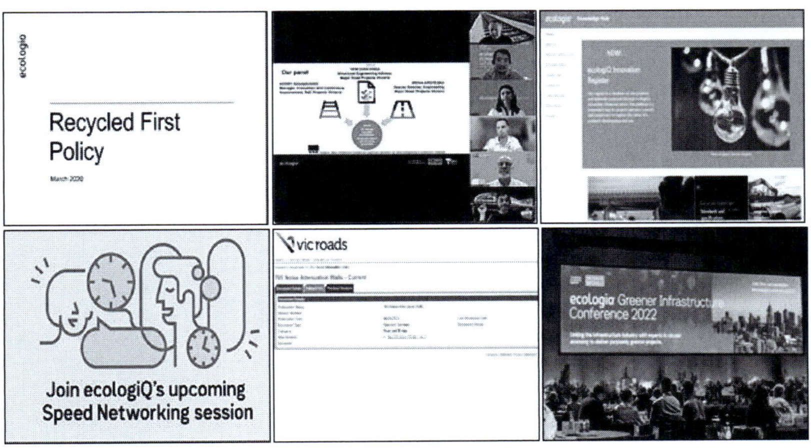

Figure 12. ecologiQ change strategies. Legend: left top (Recycled First Policy), middle top (Innovation showcase series), right top (ecologiQ Knowledge Hub, left bottom (ecologiQ Exchange), middle bottom (Reforms in PwRC specifications), right bottom (Greener Infrastructure Conference 2024).

Reforms in recycled materials specifications – ecologiQ works closely with the Department of Transport and Planning to review and change the approach to technical standards and specifications for PwRC, as well as identify priority materials for specification changes and type approval. Some of the new applications include recycled plastic products in noise walls, composite sleepers, drainage pipes, and asphalt and crumb rubber on heavily trafficked roads.

Greener Infrastructure conference – The conference explores the role of infrastructure in driving a circular economy, starting with PwRC, and the economic and resource factors shaping the industry's greener future. The conference program features diverse and esteemed keynote speakers, networking events, site tours and a dynamic trade hall where relevant suppliers showcase their products.

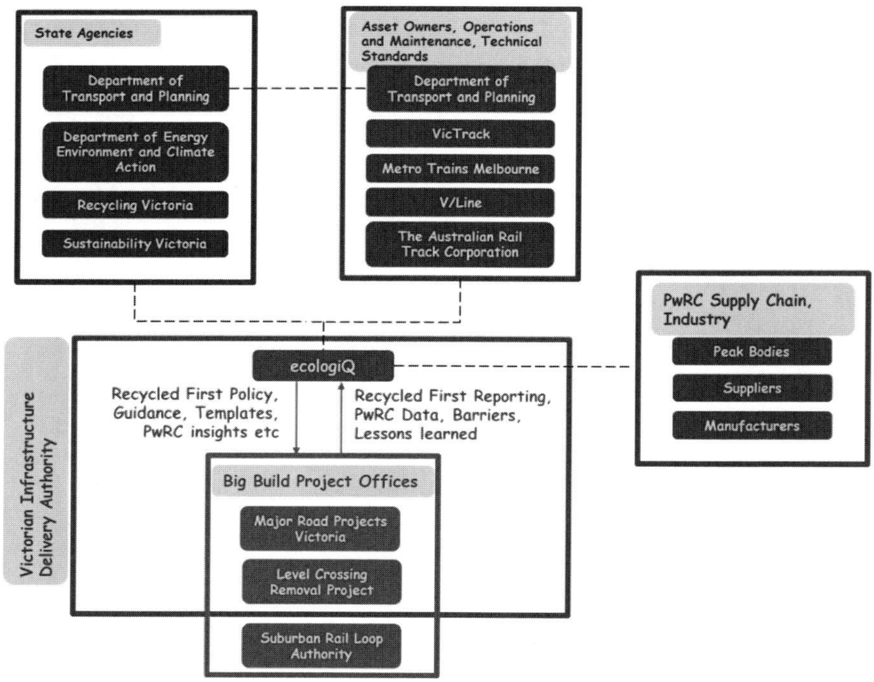

Source: Authors.

Figure 13. ecologiQ engagement model.

5.3. ecologiQ Engagement Model

This section of the chapter explores the engagement model of ecologiQ with the state agencies in connection to the change strategies mentioned in the previous section. As outlined in Figure 13, ecologiQ currently works with nine public organisations across the state including two state departments: the Department of Transport and Planning and the Department of Energy, Environment and Climate Action. ecologiQ program is also the link between the suppliers of PwRC and these public agencies.

5.4. Recognition of ecologiQ Activities

In 2023, ecologiQ was among the winners of the Premier's Sustainability Award (Industry Leader) in the Circular Economy Innovation Category. The awards, which celebrate organisations in Victoria that lead in sustainability, highlighted ecologiQ's efforts to challenge conventional approaches and demonstrate true leadership in driving greater use of recycled materials in transport infrastructure projects.

Conclusion

While ecologiQ has made great progress towards encouraging the use of PwRC in major transport infrastructure projects, other actions can boost their impact in this area. This part of the chapter provides the recommendations to serve this purpose. While some of these recommendations can be taken on board on the program level, the other recommendations need to be implemented in collaboration with other state agencies such as Sustainability Victoria, EPA and Recycling Victoria. The recommendations include:

- Help organisations set KPIs and targets related to the circular economy and the use of PwRC
- Assign a circular economy advisor to construction projects aims to advise on the best strategies for improving the use of PwRC and reporting on the achievements
- Assist businesses that use PwRC in applying for sustainability recognition programs such as IS and UDIA

- Invite representatives from industry leaders to serve on an advisory board or as members of a steering committee
- Develop a local structure to effectively deliver the program objectives while considering the contextual conditions
- Establish a close partnership with state counterparts across the country to coordinate effective initiatives and implement policies
- Expand operations into other subsectors of the building and construction industry
- Establish a regular connection with research and education providers
- Establish a system to verify the quality of PwRC
- Communicate sources to indicate which project is using PwRC
- Install signage on projects where PwRC are used, featuring a QR code that outlines the amount and methods of PwRC utilisation.

References

[1] Laybourn, P. and Lombardi, R., The role of audited benefits in Industrial Symbiosis: The U.K. National Industrial Symbiosis Programme. *Measurement + Control*, 2007. 40: p. 244-247.

[2] Coralis. *NISP: the world's first national industrial symbiosis programme*. 2022; Available from: https://www.iema.net/articles/nisp-the-symbiotic-network.

[3] Castellet-Viciano, L., Hernández-Chover, V., Bellver-Domingo, Á. and Hernández-Sancho, F., Industrial symbiosis: a mechanism to guarantee the implementation of circular economy practices. *Sustainability*, 2022. 14(23): p. 15872.

[4] PACE, *Platform for accelerating the circular economy*. 2024, PACE: Netherlands.

[5] Shooshtarian, S., Maqsood, T., Wong, P. S., Yang, R. J. and Khalfan, M., Review of waste strategy documents in Australia: Analysis of strategies for construction and demolition waste. *International Journal of Environmental Technology and Management*, 2020. 23(1): p. 1-21.

[6] Waste Authority. *Action Plans*. 2023; Available from: https://bit.ly/3yNQc6R

[7] Government of Western Australia, *Roads to reuse product specifications*. 2021.

[8] Waste Authority. *Roads to Reuse Local Government Incentives Program*. 2024; Available from: https://bit.ly/3MpMsLV

[9] Waste Authority. *Roads to Reuse: Promoting the use of recycled construction and demolition (C&D) products*. 2024; Available from: https://bit.ly/3yTE44d

[10] Waste Authority, *Waste Data Strategy*. 2019: Perth, Australia.

[11] Western Australia Government. *Finalists announced for 2024*. 2024; Available from: https://bit.ly/3X7mx0k.

[12] Local Government Areas Western Australia. *WasteSorted: a toolkit for local governments and regional councils* 2023; Available from: https://bit.ly/4g3MvdJ.

[13] Waste Authority. *WasteSorted Schools*. 2023; Available from: https://www.wasteauthority.wa.gov.au/wss/.

[14] Government of South Australia, *Green Industries SA Act 2004*, G.I. SA, Editor. 2024.

[15] GISA, *Circular Economy Knowledge Hub*. 2024, Green Industries SA Adelaide, Australia.

[16] GISA, *Green Industries Strategic Plan 2021-2025*. 2021.

[17] ARRB, *Development of an Assessment and Implementation Framework for the Use of Recycled Materials in South Australian Road Network Project*. 2022, Australian Road Research Board.

[18] GISA. *Circular opportunities in civil construction checklist*. 2022; Available from: https://bit.ly/3AD4oQq

[19] GISA. *Waste and Recycling Procurement and Contracts*. 2024; Available from: https://bit.ly/4cLIp7s

[20] GISA, *Circular impact accelerator*, G.I. SA, Editor. 2024: Adelaide, Australia.

[21] GISA. *Business Sustainability Program*. 2024; Available from: https://bit.ly/4cIhk58

[22] Sustainability Victoria, *Strategic Plan 2024-2027*. 2024.

[23] Sustainability Victoria, *ResourceSmart Schools modules*. 2024.

[24] Sustainability Victoria, *Recycled products and materials procurement toolkit for local government*. 2022.

[25] Sustainability Victoria, *Assistance with investing in Victoria's emerging Circular economy*. 2023.

[26] NSW Government, *Remanufacture NSW*. 2018.

[27] LGNSW, *Sustainable procurement guide for local government in NSW*. 2017, Local Governments NSW: Sydney, Australia.

[28] Victoria Big Build. *About ecologiQ*. 2023; Available from: https://shorturl.at/xEHM0.

[29] Sustainability Victoria. *ecologiQ: 2023 Winner – Circular economy innovation*. 2023; Available from: https://shorturl.at/gswIM.

[30] ecologiQ, *Recycled First Policy: Purposely Greener Infrastructure*, V. B. Build, Editor. 2020: Melbourne, Australia.

[31] DELWP, *Recycling Victoria: A New Economy*, L. Victoria Department of Environment, Water and Planning Editor. 2020.

Chapter 5

Training Package Development for Construction Designers

Tim Ryley[1,2]
Chamari Jayarathna[2]
and Savindi Caldera[3]

[1]School of Engineering and Built Environment, Griffith University, Brisbane, Australia
[2]Cities Research Institute, Griffith University, Brisbane, Australia
[3]School of Science, Technology & Engineering, University of the Sunshine Coast, Petrie, Australia

Abstract

This Chapter concerns the development of a training package for project designers to change their attitudes and behaviour towards the use of products with recycled content (PwRC). A systematic literature review was initially used to develop a training framework, covering both theoretical and empirical research outputs. From this framework, a training package was designed and evaluated in two training sessions. The package targeted architecture and design (A&D) community. Findings revealed that tailored material with leading edge content on PwRC works best for A&D community in the building and construction sector.

Keywords: training, construction designer, products with recycled content, construction and demolition waste

In: Policy, Practice and Education of the Circular Economy in Built Environments
Editors: Salman Shooshtarian, Peter SP Wong and Tayyab Maqsood
ISBN: 979-8-89530-623-9

1. Introduction

With the rapid growth of construction activities worldwide, which is expected to grow to a value of $15 trillion by 2030, the building and construction sector is facing a significant challenge to implement sustainable waste management due to high waste generation throughout the world [1]. On the other hand, this sector is the largest emitter of greenhouse gases accounting for 37% of global emissions due to its dependency on the high-embodied and energy intense materials such as cement, steel, and aluminium and wastes the sector generates [2]. In the United States, approximately 600 million tonnes (Mt) of construction and demolition (C&D) waste was generated in 2018 [3]. In the UK, about 137.8 Mt of C&D waste was generated in 2018 [4], and the National Waste and Resource Recovery Report 2024 estimated that Australia generated 29 Mt of C&D waste during 2022-2023 [5]. Of the global C&D waste (30% of total waste generation), a significant amount is sent to landfills irrespective of its high potential for reusing or recycling, endangering human health and the ecosystem [4]. Besides the fact that C&D waste has negative impacts, the sector also consumes scarce natural resources and generates pollution through construction operations making the situation worse. As a result of these adverse effects, sustainable waste management in the building and construction sector is of greater concern to researchers, government officials, and policymakers.

Applying the circular economy (CE) principles to the construction sector is feasible and has been investigated by many researchers [4, 6, 7]. Among the several CE strategies proposed under popular waste management models such as waste hierarchy model [8] or R-Ladder [9], the use of products with recycled content (PwRC) has been suggested as a strategy for sustainable waste management in the construction sector [10]. The researchers argue that the use of PwRC contributes to economic growth and environmental sustainability through waste reduction, energy savings, and resource conservation while promoting responsible consumption and production [10]. These contributions are aligned with CE principles of designing out waste, keeping products and materials in use, and regenerating natural systems [6].

Despite the feasibility of the CE in waste reduction and resource management in construction, there is limited attention on an individual's recycling behaviour towards C&D waste management. Changing individuals' recycling attitudes and behaviour is crucial in achieving sustainable C&D waste management, yet it has still been under-researched [11]. Some researchers have investigated the recycling behaviour of various stakeholders such as representatives from organisations operating in waste management and resource recovery sector, environmental consultants, the construction industry experts, contractors and government officials [11, 12]. However, limited research has focused on designers' attitudes and behaviour towards the use of PwRC in the building and construction sector [13, 14].

Project designers play a crucial role in influencing the use of PwRC in the construction sector through several key actions [15]. These actions include material specification prioritising PwRC, providing innovative design solutions that incorporate PwRC, educating clients and stakeholders about the benefits of using PwRC, complying with environmental standards and certifications which often require or encourage project designers to use PwRC, developing prototypes and testing them to prove the reliability and effectiveness of using PwRC in construction projects. However, a study conducted in the UK [16]. has shown that lack of information about the quality of PwRC, market availability, high cost, and negative perceptions of clients hinder project designers from playing this crucial role in influencing the use of PwRC.

Human factors have been shown to be one of the barriers in implementing waste management practices such as use of PwRC and they suggested attitudinal approaches require improvement in this context [17, 18]. Tam and Hao [19] identified that attitudes have a significant influence on construction waste recycling on site. Identifying the influence of people' attitudes and behaviour on waste recycling is challenging as they are highly subjective [20]. Bakshan, Srour [21] identified behaviour is influenced by both personal and corporate factors. However, they further pointed out that in C&D waste management, the behaviour is more sensitive to changes in personal factors such as attitude compared to corporate factors such as training [21]. Early research evidenced that intentions to increase the level of recycling behaviour are greatly influenced by non-monetary motives such as attitudes and behavioural patterns [22].

C&D waste can be reduced at both the design or planning and construction stages. Approximately 33% of materials are wasted due to designers' failure to design out waste [23, 24]. Even though designers play a critical role in C&D waste, only a small number of studies have focused on designers' behaviour towards C&D waste by design [23, 25, 26]. Researchers stated designers' perceptions of C&D waste are generally passive. For example, Osmani, Glass [23] assumed that construction waste is rarely produced at the design stage and mainly generated during site operations, Poon, Yu [27] stated that designers generally do not prioritise waste minimisation during the design process. Previous research identified several factors that hinder to adoption of waste minimisation by designers. These factors include a lack of interest [23], and a lack of experience and training [28]. Li, Tam [24] stated that most of the designers adopted a self-study approach to educating and C&D waste management. Therefore, formal training and education are required to increase the capacity of designers in waste management by design.

2. Research Objective

It has been shown that the CE rise has driven environmental improvements in the building and construction sector worldwide. Using PwRC has emerged as a targeted strategy to reduce C&D waste disposal and promote resource management. Despite the feasibility of CE in C&D waste management, changing project designer attitudes and behaviour towards the use of PwRC is challenging and poorly investigated on a global scale. Therefore, the primary research objective of this Chapter is to develop a training package which can influence project designer attitudes and behaviour changes to foster the uptake of PwRC. It covers a research approach through a systematic literature review of existing studies on the factors which influence the attitudes and beahviour change of project designers. A framework emerged from the review which applied to a training package for project designers towards the use of PwRC and implement in two industry sessions. The findings from these sessions are discussed at the end of this Chapter.

3. Theoretical Background for Pro-Environmental Behaviour Training Package Material

A myriad of theoretical frameworks have been used in explaining human behaviour across different fields such as psychology, sociology, anthropology, and economics [29]. However, the use of theory in promoting pro-environmental behaviours is limited in comparison with the other sectors such as public health [30]. Pro-environmental behaviour refers to behaviour that improves environmental conditions while acting on a reduction of negative environmental impacts. A recent review by Tian and Liu [31] identified theories related to pro-environmental behaviour based on three categories:

psychological, sociological, and economic. The review suggests that the most applied psychological theories in pro-environmental behaviour are the Theory of Planned Behaviour (TPB), Norm Activation Theory (NAT) [32], and Value-Belief-Norm theory [31].

According to the TPB, pro-environmental behaviours are the result of rational choices made expecting to maximise personal benefits [33]. This theory is centred on an individual's intention to perform a certain behaviour and the intentions can be predicted from attitudes, subjective norms, and perceived behaviour control [34]. This theory has been tested in the domain of circular economy, but the results of these studies have been mixed. While some studies suggest attitude [24], subjective norm [35], and perceived behaviour control [36] are the most important predictor of waste recycling behaviour intention, others point to environmental awareness [37]. These outcomes challenge the generalized applicability of this theory as there is a potential indirect influence of external factors such as demographic and socio-economic effects on pro-environmental behaviour [30]. Bamberg [33] also points out another scepticism against TPB, that only a few intentions translate into behaviours.

According to NAT, individuals have pro-environmental behaviour based on their moral norms which can be influenced by the awareness of consequences and the ascription of responsibility [38]. This theory provides a broad understanding of an individual's pro-environmental behaviour by emphasizing the importance of both cognitive processes and normative considerations [32, 39]. Value Belief Norm theory is another popular moral theory in investigating pro-environmental behaviour [40, 41]. It examines normative factors linked with personality elements, beliefs, and moral obligation to promote pro-environmental behaviour [42]. According to this theory, pro-environmental behaviours are more likely to appear when people value social well-being over personal interest. Even though this theory has been empirically tested in different contexts, examples from the circular economy domain are limited [30].

In sociological theories, pro-environmental behaviour is identified based on the influence of social situational factors on individual behaviour [31]. Within sociology, Social Interaction Theory (SIT) has become influential as individuals are influenced by the social network when they make behavioural decisions [43]. Social interactions provide useful information on understanding environmental issues and improve knowledge level on how to mitigate environmental issues which promotes pro-environmental behaviour [44]. In economic theories, scholars have focused on external factors such as price and income when studying pro-environmental behaviour [31]. Assumptions in economics are that people behave rationally and maximise personal utility. Under this approach, incentives are considered an effective way to promote environmental behaviour [45]. In economic models, incentives and an individual's willingness to pay for environmental improvements have been widely used as determinants of pro-environmental behaviour [46, 47].

In addition to behavioural theories discussed under psychology, sociology and economics, various intervention strategies have been used to understand pro-environmental behaviour. Two popular intervention strategies are nudging, and community-based social marketing (CBSM), which are based on cognitive biases and social influences respectively [48, 49]. In nudging, it is assumed that humans are not exclusively rational beings and do not always behave based on their knowledge and intentions [49]. Even previous research has used nudges in environmental research such as reducing food waste and plastic waste [50, 51], promoting green energy and more durable products [52, 53], a thorough methodological approach is still required to decide a right nudging strategy as not all nudging strategies are equally effective in controlling a particular behaviour [54].

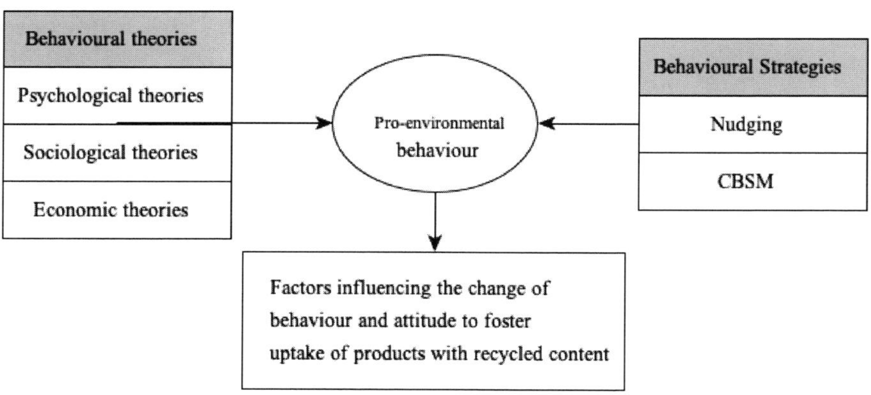

Source: Authors.

Figure 1. A proposed theoretical framework.

CBSM strategies use marketing principles and techniques to influence human behaviour that benefits the individual and society [55]. Compared to other theories and frameworks, CBSM has more options for behavioural changes. CBSM tools include communication, incentives, norming, and social diffusion [30]. Even though several environmental studies used CBSM [56], the use of CBSM in the circular economy is limited. Based on this background, we propose a theoretical framework (Figure 1) behind the pro-environmental behaviour which leads to identifying the factors that influence the project designers to change their attitudes and behaviour towards the use of PwRC in construction projects.

4. Theoretical Review: Discussion

This section provides discussion from a systematic literature review of theory relating to different factors (including strategies and barriers) that influence the attitudes and behaviour change of project designers towards the use of PwRC.

4.1. Psychological Factors that Affect the Use of PwRC

The most recognised psychological factor that affects the use of PwRC is attitude. According to TPB, attitude refers to the "degree to which a person has a favourable or unfavourable evaluation or appraisal of the behaviour" [34]. Our analysis shows that the attitude towards the use of PwRC is mainly determined by factors such as perceived cost, certification of PwRC, negative perception of clients, perceived benefit, and perceived risk and quality. Several studies have shown that designers perceive the cost of PwRC to be higher than virgin products [16, 57, 58]. Perceived costs include those for sorting and collection, transport and storing, and quality control and assurance of PwRC. Jain, Singhal [57] suggested that perceived cost negatively influences an individual's attitude towards recycling. Therefore, perceived cost is a main determinant of the designer's attitude towards the use of PwRC.

The certification process has a high impact in terms of the acceptance of recycled aggregates [59]. Certification on recycled aggregates give project designers the confidence to use them in their projects. Thus, it is important to establish a certification program for recycled aggregates in order to change attitudes [14]. Ramos, Martinho [12] also emphasise the use of specific standards for recycled aggregates to comply with. Specific standards give designers a benchmark to positively think about the use of recycled aggregates in their projects. Designers rarely specify recycled products following a negative perception of clients [12, 16, 59]. Apart from designers who are responsible for the material specification, clients play a major role in the type of materials required for their projects. Generally, clients consider recycled products to be lower quality, thus they have a negative perception of the use of recycled products in their projects. The perceived risk of using recycled products also caused to client's negative perception of PwRC [12]. This means that client perceptions can sometimes determine designer attitudes towards the use of PwRC.

Perceived benefit or usefulness is another factor that determines the attitude. Designers are hesitant to use PwRC unless they realise the benefits. These benefits can include environmental [11], corporate image or commercial acquisition [20]. Literature suggests that Perceived benefit/usefulness has a positive attitude towards construction waste management [36, 57], thus it influences the designer's attitudes towards the use of PwRC. A lack of confidence in quality or perceived quality also negatively affects the designer's attitudes towards the use of PwRC [57].

Subjective norms are social factors that affect individuals to perform a certain behaviour [34]. These social norms refer to external pressures that influence an individual's behaviour through regulations (regulatory pressure) seeking peer opinion (normative pressure) or what others in a similar situation follow or not (mimetic pressure). These norms are aligned with Institutional Theory. In addition to these three pressures, cultural beliefs can also be identified under subjective norms. According to our analysis, many of the authors identify regulatory pressure [11, 57], and normative pressure [59] as the most influential subjective norms in changing behaviour towards the use of PwRC.

Ramos and Martinho [59] highlighted that environmental policies that promote the use of recycled aggregates are important in enhancing the confidence of using it. This is also the case for legal specifications. Recommendations from peers are also important in changing behaviour towards the use of PwRC [12]. However, compared to attitude and perceived behaviour control, subjective norms received low recognition in changing behaviour towards PwRC. Li, Tam [24] also state that subjective norms play a subordinate role in designers' intention towards construction waste management while Jain, Singhal [57] found no significant relationship between subjective norms and recycling behaviour.

Perceived behaviour control refers to factors that control a particular behaviour [34]. It embodies the concept of "ipsative behaviour" proposed by Frey [60] as mentioned by Jain, Singhal [57]. Ipsative behaviour suggests that perceived internal cognitive and external constraints govern the individual's particular behaviour [60]. According to our analysis, the level of education and training is the most refereed internal control factor towards behaviour change. Li, Tam [24] highlighted that the level of education of designers may have an impact on their behaviour towards construction waste minimization by design.

Therefore, Mohsen and Matarneh [14] emphasised the importance of integrating sustainability principles into the curriculum of project design programs and professional development courses, educating clients and users about the positive impact of sustainable design choices, and developing and disseminating case studies of successful applications of sustainability practices. However, Olanrewaju and Ogunmakinde [13] stated that even though architects are expected to go through formal education, they may lack detailed training on waste minimisation. This could lead to other internal control factors such as a lack of confidence [12], a lack of experience, a lack of expertise in potential applications of PwRC [10] and resistance to new technology [61]. Researchers highlighted that professional experience is important in changing the attitudes and behaviour of PwRC users [4, 12, 21]. The most recognised external control factor is a lack of government support. This could include introducing mandatory requirements and financial incentives for the use of PwRC in construction projects [62]. Other external control factors are market availability [12, 63, 64], work environment [20, 65] and a lack of information about PwRC [16, 64].

According to NAT, moral norms towards pro-environmental behaviour can be influenced by the awareness of consequences and the ascription of responsibility [38]. Even though the moral norms were initially introduced based on the NAT, recent researchers attempted to include them under Extended Theory of Planned Behaviour (ETPB). Chan and Bishop [66] found that empirical support for moral norms should be included in a TPB model. The rationale behind this inclusion is that recycling behaviour can be best conceptualized as a moral rather than economic behaviour as it describes what behaviour is right or wrong [66]. Thus, a sense of responsibility in terms of moral obligation [65] environmental consciousness [57], and environmental awareness [21, 65] are considered moral norms in changing behaviour towards the use of PwRC according to NAT.

Knowledge is one of the most recognized psychological factors towards the use of PwRC. In fact, researchers identified knowledge as an extended variable in the TPB model [36]. Knowledge-related factors contributing to low uptake of PwRC include a lack of expertise and understanding of PwRC utilisation [67], a lack of market survey [13], and insufficient training at academic and professional levels [68]. Market surveys and training are good sources of knowledge for project designers or architects to understand available PwRC and their applications.

4.2. Economic Factors that Affect the Use of PwRC

According to economists, people are rational and try to maximise their personal utility [31]. Under this logic, internal and external motivations are considered in changing environmental behaviour. These motivations include economic benefits such as financial incentives [58] reward schemes [26] competitive advantage [67] and environmental benefits such as reducing virgin material extraction, waste disposal and carbon emissions [67]. Of these economic factors, financial incentives are highly considered in terms of motivation to change behaviour [62]. A high cost of circular materials, a high investment cost, and a low return on investment are some of the greater challenges ahead of adopting circular design in construction projects. The use of PwRC is essential for circular design, thus financial incentives such as tax rebates, subsidies on import of PwRC, statutory fees and circular building design are suggested to motivate project designers towards the use of PwRC [68].

4.3. Sociological Factors that Affect the Use of PwRC

According to the SIT, individuals are influenced by their social network when they make behavioural changes [43]. Based on SIT, authors identified collaboration and communication as sociological factors that influence the designers to change their attitudes and behaviour towards PwRC. Mohsen and Matarneh [14] emphasised the importance of interdisciplinary collaboration in the design stage for the uptake of circular design through PwRC. Collaboration with universities and research institutions can alleviate the challenges of using PwRC [64]. Researchers highlighted that training could facilitate collaboration among project designers to develop technical skills through knowledge sharing [68]. Communication is an effective tool to improve social interaction towards changing behaviour. Communication with other professionals, government officials, and construction enterprises can influence project designers to change their attitude towards the use of PwRC [13, 64].

Social interactions through collaboration and communication provide useful information to make decisions on pro-environmental behaviour [44].

4.4. Strategies and Barriers to Change the Attitudes and Behaviour Towards the Use of PwRC

Most researchers identified training as an effective strategy for construction waste management in broad terms [62, 69]. Other frequently mentioned strategies are education and government support [70, 71]. In addition, researchers identified market surveys, project management planning, technology, and expanding product types and markets as strategies to change the behaviour towards the use of PwRC (see Table 1). From the perspectives of barriers, major factors that influence the use of PwRC are high cost, a lack of awareness about PwRC, and a lack of government support. A few studies identified a lack of training [13, 61], lack of recycling facility [58, 72], and cultural belief [68, 72].

Most of the strategies and barriers are complementing each other and they impact the behaviour through psychological, economic, or sociological factors. For example, behaviour towards the use of PwRC can be changed through perceived behaviour control factors such as government support, lack of recycling facility, knowledge related factors such as level of training, and education, subjective norms such as cultural belief and regulations, economic factors such as financial benefits. Lockrey, Nguyen [70] stated that education has a high potential for changing behaviour in planning, tracking and utilization of concrete waste. Government support can include introducing relevant policies such as waste minimisation policy [10] financial subsidies and tax reductions [62] to encourage construction companies to use PwRC which in turn positively influences the change in the behaviour of project designers too [63]. Barriers such as lack of awareness, lack of training, lack of experience, and lack of education can be avoided through strategies such as proper training, and education emphasising the benefits of using PwRC, the application of PwRC [68]. A strategy like a market survey provides information for project designers to understand available construction materials and their functionality [13].

Table 1. Factors (a), Strategies (b) and Barriers (c) influencing the change of attitudes and behaviours

(a) Factors that influence the behaviour	How these factors influence the behaviour
Perceived value	(+) impact through attitude
Perceived usefulness/benefit	(+) impact through attitude
perceived risk	(-) impact through attitude
Perceived quality	(+) impact through attitude
Perceived cost	(-) impact through attitude
Negative perception of clients	(-) impact through attitude
Employer's support	(+) impact through attitude
Employees' perception	(+/-) impact through attitude
Certification	(+) impact through attitude
Normative pressure	(+) impact through subjective norms
Regulatory pressure	(+) impact through subjective norms
Mimetic pressure	(+) impact through subjective norms
Level of education	(+) impact through perceived behaviour control
Experience	(+) impact through perceived behaviour control
Design capability	(-) impact through perceived behaviour control
Lack of confidence	(-) impact through perceived behaviour control
Lack of information	(-) impact through perceived behaviour control
Market availability	(-) impact through perceived behaviour control
Opportunity	(-) impact through perceived behaviour control
Physical capability	(-) impact through perceived behaviour control
Physical stress	(-) impact through perceived behaviour control
Psychological stress	(-) impact through perceived behaviour control
Supervision	(-) impact through perceived behaviour control
Training	(+) impact through perceived behaviour control
Tedious operational process	(+) impact through perceived behaviour control
Organizational size	(-) impact through perceived behaviour control
Environmental awareness	(-) impact through moral norms
Environmental consciousness	(+) impact through moral norms
Sense of responsibility	(+) impact through moral norms
Knowledge	(+) impact through perceived behaviour control
Economic benefits	(+) impact through perceived behaviour control
Competitive advantage	(+) impact through motivations
Environmental benefits	(+) impact through motivations
Communication	(+) impact through motivations
Collaboration	(+) impact through social network
(b) Strategies	How these strategies influence the behaviour
Training	(+) impact through perceived behaviour control
Education	(+) impact through perceived behaviour control
Government support	(+) impact through perceived behaviour control
Market survey	(+) impact through knowledge
Project management planning	(+) impact through perceived behaviour control
Technology	(+) impact through perceived behaviour control
Expanding product types and market	(+) impact through perceived behaviour control
Financial incentives	(+) impact through motivations

(b) Strategies	How these strategies influence the behaviour
Stakeholders support	(+) impact through perceived behaviour control
Waste minimization policy	(+) impact through subjective norms
(c) Barriers	How these barriers influence the behaviour
High cost	(-) impact through attitude
Lack of awareness	(-) impact through perceived behaviour control
Lack of government support	(-) impact through perceived behaviour control
Cultural beliefs	(-) impact through perceived behaviour control
Lack of recycling facility	(-) impact through perceived behaviour control
Lack of training	(-) impact through perceived behaviour control
Design teams' poor experience	(-) impact through perceived behaviour control
Inconsistency of product quality	(-) impact through attitude
Lack of coordination between demand and supply	(-) impact through perceived behaviour control
Lack of expertise and understanding of PwRC use	(-) impact through perceived behaviour control
Lack of marketing strategy	(-) impact through attitude
Lack of product competitiveness	(-) impact through attitude
Lack of second-hand market	(-) impact through perceived behaviour control
Lack of waste classification	(-) impact through perceived behaviour control
Resistance to new technology	(-) impact through perceived behaviour control
Lack of trust	(-) impact through attitude

5. Design and Implementation of the Training Package

This section presents a framework developed for designing a training package to enable attitudes and behavioural change for project designers towards the use of PwRC. There is then a discussion of how a training package was designed and implemented in two training sessions.

5.1. Framework for Designing a Training Package to Change the Attitudes and Behaviour of Project Designers Towards the Use of PwRC

It has been identified that training is interrelated with many of the other factors that influence the change of attitudes and behaviour. In terms of psychological factors, training was identified as a perceived behaviour (external) control factor. Some researchers identified lack of training as a barrier to changing attitudes and behaviour [13, 57], while other studies identified it as a strategy to change the behaviour and attitude towards the use of PwRC [64, 69]. Moreover, training is directly linked with factors such as knowledge, information, awareness, confidence, perceived usefulness, perceived risk, perceived usefulness/benefit, negative perception, and experience.

Therefore, appropriate training can positively impact the factors that determine the change of attitudes and behaviour which ultimately helps change attitudes and behaviour towards the use of PwRC. Previous studies also highlighted the importance of training in changing attitudes and behaviour regarding the use of PwRC [10]. With this understanding from the review results, we have introduced a framework that can be used to develop a training package to change the attitudes and behaviour of project designers towards the use of PwRC (see Figure 2). This framework illustrates the purpose of training, what kind of content should be included and how they support achieving the purpose (changing attitudes and behaviour of project designers towards the use of PwRC). The content is based on the highly referenced influential factors identified from the literature review analysis. Accordingly, we propose seven elements (environmental benefits and CE concept, cost-benefit analysis, certification and standards, examples of successful projects that have used PwRC, information on existing rules, regulations and policies, application of PwRC, circular design strategies) for inclusion in the training package that can influence in changing attitudes and behaviour of project designers towards the use of PwRC.

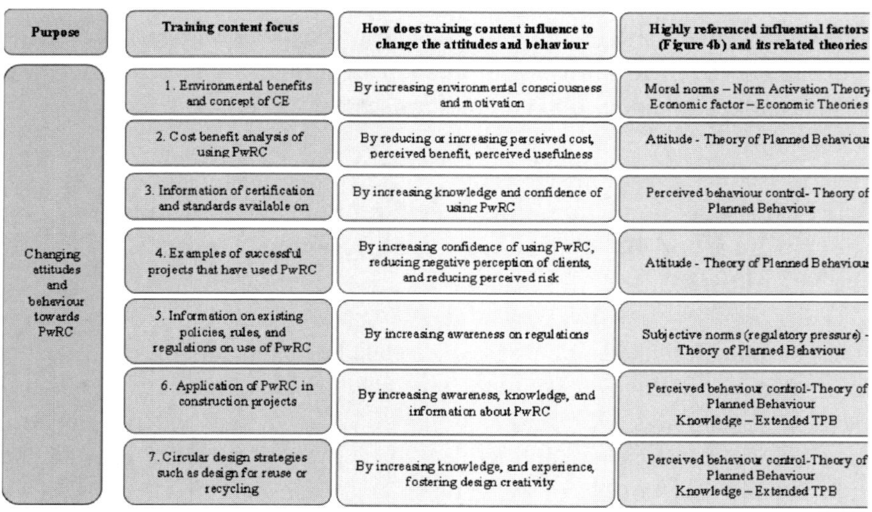

Source: Jayarathna, Ryley [73].

Figure 2. A framework for developing a training package to change attitudes and behaviour towards the use of PwRC.

5.2. Design of the Training Package Material

The focus of this training was on construction architecture and design (A&D) community, at the design phase of construction. Development Managers are also, therefore, key in this aspect as they are the decision-makers and commissioners. Aligning with the framework which emerged from the theoretical review, the seven elements were incorporated into the training packages, and organised according to the following five topics:

1. Understanding the C&D waste problem
2. Sustainability and the CE
3. The design phase and sustainable construction strategies
4. PwRC
5. Case studies tailored to the organization.

The training package content fully aligned with the framework shown in Figure 2 and the seven items listed under training content focus. The environmental benefits and concept of CE (1) were covered under the Sustainability and Circular Economy topic. Circular design strategies such as design for reuse or recycling (7) were covered until the Design Phase & Sustainable Construction Strategies topic. For the PwRC topic, this included cost benefit analysis (2), information of certification and standards available (3), examples of successful projects that have used PwRC (4), and information on existing policies, rules, and regulations on use of PwRC (5). Finally, the case studies tailored to the organisation was the application of PwRC in construction projects (6).

Some of the material tailored to the industry partner was case study material analysed during the Sustainable Built Environment National Research Centre (SBEnrc) projects. The ATCO Structures and Logistics training session contained a section on their modular housing case study. Modular construction is a building method where individual components, or modules, are prefabricated in a factory-controlled environment and then transported to the construction site. These modules are typically designed to fit together seamlessly to form a complete structure, such as a home or commercial building. The identified economic benefits of using PwRC in modular construction were minimum defects, an extended PwRC lifecycle, a standardised application, an application of a wider range of PwRC, and an opportunity to demonstrate environmental sustainability. Corresponding environmental benefits were an improved carbon footprint, keeping waste resources out of landfills, contamination reduction, reduced noise pollution and reduced ecological impacts.

The DevelopmentWA training session contained tailored material on their case study in Hamilton Hill in Perth. The project, completed in 12 months in 2019 with a budget of $2.7 million, aimed to redevelop an old high school and neighbouring land (11.9 ha) into a residential estate to provide residential opportunities for Western Australian communities. Application of PwRC in the project included the following: using salvaged timber in landscaping features; reusing 40,000 clay bricks and roof tiles as aggregates underneath the drainage infrastructure; reusing old bricks to create brick walls and a brick toilet block; reusing crushed brick, tiles, concrete and other material into the road sub-base; and using 2,425 m^3 of recycled concrete in the retaining walls. A further training package was developed of the combined material a recording was undertaken in a studio at Griffith University Nathan campus in October 2024. The material and recording are freely available on the SBEnrc website[1].

5.3. Findings from the Training Sessions

The two training sessions were undertaken in September 2024.

Training Session 1

Industry partner: ATCO Structures and Logistics. Overview of participants and training expectations

Location: ATCO Structures in Luscombe, Southeast Queensland.

Delivered by Prof. Tim Ryley in person; Dr Salman Shooshtarian joined online.

Participants: 16 in room and 1 online. Most attendees completed the feedback form.

Content: 35 slide PPT presentation.

Most of the ATCO attendees were in technical or professional positions, particularly in a Designer pr Estimator role. Some had management or administrative roles. There were 13 attendees who completed a post-session

[1] A copy of the training package is available at: https://sbenrc.com.au/environment/1-95/.

survey. The most useful information for attendees was the details around the PwRC. For training needs moving forward, the most common response was on new materials and methods, and then how they could be applied to ATCO Structures. This demonstrates the importance of providing leading-edge material which is then tailed to the industry partner. Most attendees prefer training in-person, although some also wanted a recording afterwards.

Training Session 2

Industry partner: DevelopmentWA. Overview of participants and training expectations

Location: DevelopmentWA, Perth, Western Australia.

Delivered by Prof. Tim Ryley online.

Participants: 20 online. A handful of attendees completed the feedback form.

Content: 36 slide PPT presentation.

The people attending the session from DevelopmentWA were primarily development managers. The positive elements related to the detailed information within the slides. There were four attendees who responded to the post-session survey. For improvements, more specific details on the materials, tools and case studies would have been beneficial, again demonstrating a need to up-to-date and leading-edge material. A key outcome from the session and responses, is how local government can overcome the barriers and issues related to PwRC raised in the session.

Conclusion

Changing attitudes and behaviour towards the use of PwRC has received greater international attention along with a recent sharp increase in studies on pro-environmental behaviour. However, there are a limited number of studies focusing on training and construction project designer recycling behaviour. From this systematic literature review, changing attitudes and behaviour towards the use of PwRC is particularly influenced by the psychological factors of attitude, knowledge, perceived behaviour control and subjective norms. Moreover, the top three strategic factors identified were training, education and government support, whereas the top three barriers were high associated costs, a lack of awareness, and a lack of government support. Based on the literature review findings, a framework was developed and implemented for the training of project designers to change their attitudes and behaviour towards the use of PwRC. From the experience of running the training sessions, together with attendee feedback, it is important that up-to-date and leading-edge PwRC material is presented, and that it is tailored to the industry partner and attendees who are typically in design and management roles. The key contribution of this Chapter is to develop a framework that can be applied elsewhere for training on PwRC. This training framework provides guidance on seven key training content areas for construction and sustainability practitioners, as well as policymakers, to design tailored educational programs that promote the use of PwRC

References

[1] Zafar, S., *Sustainable waste Management in the Construction Industry*. 2021, Bio Energy Consult. Available at: https://www.bioenergyconsult.com

[2] UNEP, *Building materials and the climate: Constructing a new future*. 2023, United Nations Environment Programme, Yale Center for Ecosystems Architecture.

[3] USE EPA, *Construction & Demolition Debris: Material specific data*, in *US EPA: Research Triangle Park, NC, USA*, C.D.D.M.s. data, Editor. 2018.

[4] Ogunmakinde, O. E., Egbelakin, T. and Sher, W., Contributions of the circular economy to the UN sustainable development goals through sustainable construction. *Resources, Conservation and Recycling*, 2022. 178: p. 106023.

[5] Blue Environment, *National waste and resource recovery report 2024*, E. Department of Climate Change, the Environment and Water, Editor. 2025: Canberra, Australia.

[6] Zaman, A., Caceres Ruiz, A. M., Shooshtarian, S., Ryley, T., Caldera, S. and Maqsood, T., Development of the circular economy design guidelines for the Australian built environment sector. *Sustainability*, 2023. 15(3): p. 2500.

[7] Shooshtarian, S., Maqsood, T., Caldera, S. and Ryley, T., Transformation towards a circular economy in the Australian construction and demolition waste management system. *Sustainable Production and Consumption*, 2022. 30(2021): p. 89-106.

[8] Zhang, C., Hu, M., Di Maio, F., Sprecher, B., Yang, X. and Tukker, A., An overview of the waste hierarchy framework for analyzing the circularity in construction and demolition waste management in Europe. *Science of the Total Environment*, 2022. 803: p. 149892.

[9] Potting, J., Hekkert, M. P., Worrell, E. and Hanemaaijer, A., Circular economy: measuring innovation in the product chain. *Planbureau voor de Leefomgeving*, 2017(2544).

[10] Shooshtarian, S., Maqsood, T., Wong, P. S. P., Caldera, S., Ryley, T., Zaman, A. and Cáceres Ruiz, A. M., Circular economy in action: the application of products with recycled content in construction projects – a multiple case study approach. *Smart and Sustainable Built Environment*, 2024. 13(2): p. 370-394.

[11] Mak, T. M., et al., Extended theory of planned behaviour for promoting construction waste recycling in Hong Kong. *Waste management*, 2019. 83: p. 161-170.

[12] Ramos, M., Martinho, G., Vasconcelos, L. and Ferreira, F., Local scale dynamics to promote the sustainable management of construction and demolition waste. *Resources, Conservation & Recycling Advances*, 2023. 17: p. 200135.

[13] Olanrewaju, S. D. and Ogunmakinde, O. E., Waste minimisation strategies at the design phase: Architects' response. *Waste Management*, 2020. 118: p. 323-330.

[14] Mohsen, M. S. and Matarneh, R., Exploring the interior designers' attitudes toward sustainable interior design practices: The case of Jordan. *Sustainability*, 2023. 15(19): p. 14491.

[15] Shooshtarian, S., Caldera, S., Maqsood, T. and Ryley, T., Using recycled construction and demolition waste products: A review of stakeholders' perceptions, decisions, and motivations. *Recycling*, 2020. 5(4): p. 31.

[16] Oyedele, L. O., Ajayi, S. O. and Kadiri, K. O., Use of recycled products in UK construction industry: An empirical investigation into critical impediments and strategies for improvement. *Resources, Conservation and Recycling*, 2014. 93: p. 23-31.

[17] Udawatta, N., Zuo, J., Chiveralls, K. and Zillante, G., Improving waste management in construction projects: An Australian study. *Resources, Conservation and Recycling*, 2015. 101: p. 73-83.

[18] Udawatta, N., Zuo, J., Chiveralls, K., Yuan, H., George, Z. and Elmualim, A., Major factors impeding the implementation of waste management in Australian construction projects. *Journal of Green Building*, 2018. 13(3): p. 101-121.

[19] Tam, V. W. and Hao, J. J. *Attitudes towards recycling on construction sites*. in *Proceedings of the Institution of Civil Engineers-Waste and Resource Management*. 2016. Thomas Telford Ltd.

[20] Tam, V., Le, K., Wang, J. and Illankoon, I., Practitioners Recycling Attitude and Behaviour in the Australian Construction Industry. *Sustainability*, 2018. 10(4): p. 1-23.

[21] Bakshan, A., Srour, I., Chehab, G., El-Fadel, M. and Karaziwan, J., Behavioral determinants towards enhancing construction waste management: A Bayesian Network analysis. *Resources, Conservation and Recycling*, 2017. 117: p. 274-284.

[22] De Young, R., Recycling as appropriate behaviour. A review of survey data from selected education programmes in Michigan. Resc. *Conservation and Recycling*, 1990. 3: p. 253-266.

[23] Osmani, M., Glass, J. and Price, A. *Architect and contractor attitudes to waste minimisation*. in *Proceedings of the Institution of Civil Engineers-waste and resource management*. 2006. Thomas Telford Ltd.

[24] Li, J., Tam, V. W., Zuo, J. and Zhu, J., Designers' attitude and behaviour towards construction waste minimization by design: A study in Shenzhen, China. *Resources, Conservation and Recycling*, 2015. 105: p. 29-35.

[25] Osmani, M., Glass, J. and Price, A. D., Architects' perspectives on construction waste reduction by design. *Waste Management*, 2008. 28(7): p. 1147-1158.

[26] Wang, J., Li, Z. and Tam, V. W., Critical factors in effective construction waste minimization at the design stage: a Shenzhen case study, China. *Resources, Conservation and Recycling*, 2014. 82: p. 1-7.

[27] Poon, C. S., Yu, A. T. and Jaillon, L., Reducing building waste at construction sites in Hong Kong. *Construction Management and Economics*, 2004. 22(5): p. 461-470.

[28] Nguyen, T. P. and Chileshe, N., Revisiting the construction project failure factors in Vietnam. *Built Environment Project and Asset Management*, 2015. 5(4): p. 398-416.

[29] Davis, R., Campbell, R., Hildon, Z., Hobbs, L. and Michie, S., Theories of behaviour and behaviour change across the social and behavioural sciences: a scoping review. *Health psychology review*, 2015. 9(3): p. 323-344.

[30] Parajuly, K., Fitzpatrick, C., Muldoon, O. and Kuehr, R., Behavioral change for the circular economy: A review with focus on electronic waste management in the EU. *Resources, Conservation & Recycling:* 6(2020): p. 100035.

[31] Tian, H. and Liu, X., Pro-environmental behavior research: Theoretical progress and future directions. *International Journal of Environmental Research and Public Health*, 2022. 19(11): p. 6721.

[32] Klöckner, C. A., A comprehensive model of the psychology of environmental behaviour - A meta-analysis. *Global Environmental Change*, 2013. 23(5): p. 1028-1038.

[33] Bamberg, S., Changing environmentally harmful behaviors: A stage model of self-regulated behavioral change. *Journal of Environmental Psychology*, 2013. 34: p. 151-159.

[34] Ajzen, I., *The theory of planned behaviour: Reactions and reflections*. 2011, Taylor & Francis. p. 1113-1127.

[35] Echegaray, F. and Hansstein, F. V., Assessing the intention-behavior gap in electronic waste recycling: the case of Brazil. *Journal of Cleaner Production*, 2017. 142: p. 180-190.

[36] Swarna Swetha, K., Tezeswi T. P. and Siva Kumar M. V. N. Implementing construction waste management in India: An extended theory of planned behaviour approach. *Environmental Technology & Innovation*, 2022. 27: p. 102401.

[37] Wang, Z., Guo, D. and Wang, X., Determinants of residents'e-waste recycling behaviour intentions: evidence from China. *Journal of Cleaner Production*, 2016. 137: p. 850-860.

[38] Schwartz, S. H., *Normative influences on altruism*, in *Advances in experimental social psychology*. 1977, Elsevier. p. 221-279.

[39] Oh, J. and Ki, E. J., Extending norm activation theory to understand publics' support for environmentally responsible organizations. *Corporate Communications: An International Journal*, 2023. 28(3): p. 381-399.

[40] Janmaimool, P. and Denpaiboon, C., Evaluating determinants of rural Villagers' engagement in conservation and waste management behaviors based on integrated conceptual framework of Pro-environmental behavior. *Life sciences, society and policy*, 2016. 12: p. 1-20.

[41] Bronfman, N. C., Cisternas, P. C., López-Vázquez, E., De la Maza, C. and Oyanedel, J. C., Understanding attitudes and pro-environmental behaviors in a Chilean community. *Sustainability*, 2015. 7(10): p. 14133-14152.

[42] Stern, P. C., New environmental theories: toward a coherent theory of environmentally significant behavior. *Journal of Social Issues*, 2000. 56(3): p. 407-424.

[43] Kim, J., Bae, Y. K. and Chung, J. H., Modeling social distance and activity-travel decision similarity to identify influential agents in social networks and geographic space and its application to travel mode choice analysis. *Transportation Research Record*, 2020. 2674(6): p. 466-479.

[44] Yin, J. and Shi, S., Social interaction and the formation of residents' low-carbon consumption behaviors: An embeddedness perspective. *Resources, Conservation and Recycling*, 2021. 164: p. 105116.

[45] Clark, C. F., Kotchen, M. J. and Moore, M. R., Internal and external influences on pro-environmental behavior: Participation in a green electricity program. *Journal of Environmental Psychology*, 2003. 23(3): p. 237-246.

[46] Nnorom, I., Ohakwe, J. and Osibanjo, O., Survey of willingness of residents to participate in electronic waste recycling in Nigeria–A case study of mobile phone recycling. *Journal of Cleaner Production*, 2009. 17(18): p. 1629-1637.

[47] Atlason, R. S., Giacalone, D. and Parajuly, K., Product design in the circular economy: Users' perception of end-of-life scenarios for electrical and electronic appliances. *Journal of Cleaner Production*, 2017. 168: p. 1059-1069.

[48] McKenzie-Mohr, D., New ways to promote proenvironmental behavior: Promoting sustainable behavior: An introduction to community-based social marketing. *Journal of Social Issues*, 2000. 56(3): p. 543-554.

[49] Thaler, R. and Sunstein, C. *Nudge: Improving decisions about health, wealth and happiness*. in *Amsterdam Law Forum; HeinOnline: Online*. 2008. HeinOnline.

[50] Kallbekken, S. and Sælen, H., 'Nudging'hotel guests to reduce food waste as a win–win environmental measure. *Economics letters*, 2013. 119(3): p. 325-327.

[51] Rivers, N., Shenstone-Harris, S. and Young, N., Using nudges to reduce waste? The case of Toronto's plastic bag levy. *Journal of Environmental Management*, 2017. 188: p. 153-162.

[52] Ebeling, F. and Lotz, S., Domestic uptake of green energy promoted by opt-out tariffs. *Nature Climate Change*, 2015. 5(9): p. 868-871.

[53] European Commission, *Behavioural study on consumers' engagement in the circular economy*. 2018. European Commission: Consumers, Health, Agriculture and Food Executive Agency

[54] Momsen, K. and Stoerk, T., From intention to action: Can nudges help consumers to choose renewable energy?. *Energy Policy*, 2014. 74: p. 376-382.

[55] Salazar, G., Mills, M. and Veríssimo, D., Qualitative impact evaluation of a social marketing campaign for conservation. *Conservation Biology*, 2019. 33(3): p. 634-644.

[56] Cole, E. J. and Fieselman, L., A community-based social marketing campaign at Pacific University Oregon: Recycling, paper reduction, and environmentally

preferable purchasing. *International Journal of Sustainability in Higher Education*, 2013. 14(2): p. 176-195.

[57] Jain, S., Singhal, S., Jain, N. K. and Bhaskar, K., Construction and demolition waste recycling: Investigating the role of theory of planned behavior, institutional pressures and environmental consciousness. *Journal of Cleaner Production*, 2020. 263: p. 121405.

[58] Dokter, G., Thuvander, L. and Rahe, U., How circular is current design practice? Investigating perspectives across industrial design and architecture in the transition towards a circular economy. *Sustainable Production and Consumption*, 2021. 26: p. 692-708.

[59] Ramos, M. and Martinho, G., Relation between construction company size and the use of recycled materials. *Journal of Building Engineering*, 2021: p. 103523.

[60] Frey, B. S., Ipsative and objective limits to human behavior. *Journal of Behavioral Economics*, 1988. 17(4): p. 229-248.

[61] Kineber, A. F., Massoud, M. M., Hamed, M. M. and Qaralleh, T. J. O., Exploring sustainable interior design implementation barriers: A partial least structural equation modeling approach. *Sustainability*, 2023. 15(5): p. 4663.

[62] Luciano, A., Cutaia, L., Altamura, P. and Penalvo, E., Critical issues hindering a widespread construction and demolition waste (CDW) recycling practice in EU countries and actions to undertake: the stakeholder's perspective. *Sustainable Chemistry and Pharmacy*, 2022. 29: p. 100745.

[63] Ma, M., Tam, V. W., Le, K. N. and Li, W., Challenges in current construction and demolition waste recycling: A China study. *Waste Management*, 2020. 118: p. 610-625.

[64] Ding, Z., Wang, X. and Zou, P. X., Barriers and countermeasures of construction and demolition waste recycling enterprises under circular economy. *Journal of Cleaner Production*, 2023. 420: p. 138235.

[65] Hao, J. L., Yu, S., Tang, X. and Wu, W., Determinants of workers' pro-environmental behaviour towards enhancing construction waste management: Contributing to China's circular economy. *Journal of Cleaner Production*, 2022. 369: p. 133265.

[66] Chan, L. and Bishop, B., A moral basis for recycling: Extending the theory of planned behaviour. *Journal of Environmental Psychology*, 2013. 36: p. 96-102.

[67] Shooshtarian, S., Maqsood, T., Wong, P. S., Caldera, S., Ryley, T., Zaman, A. and Ruiz, A., Circular economy in action: The application of products with recycled content in construction projects: A multiple case study approach. *Smart and Sustainable Built Environment*, 2024. 13(2): p. 370-394.

[68] Suleman, T., Ezema, I. and Aderonmu, P., Challenges of circular design adoption in the Nigerian built environment: An empirical study. *Cleaner Engineering and Technology*, 2023. 17: p. 100686.

[69] Akinade, O. O., et al., Design for Deconstruction (DfD): Critical success factors for diverting end-of-life waste from landfills. *Waste Management*, 2017. 60: p. 3-13.

[70] Lockrey, S., Nguyen, H., Crossin, E. and Verghese, K., Recycling the construction and demolition waste in Vietnam: opportunities and challenges in practice. *Journal of Cleaner Production*, 2016. 133: p. 757-766.

[71] Ganiyu, S. A., Oyedele, L. O., Akinade, O., Owolabi, H., Akanbi, L. and Gbadamosi, A., BIM competencies for delivering waste-efficient building projects in a circular economy. *Developments in the Built Environment*, 2020. 4: p. 100036.

[72] Charef, R., Ganjian, E. and Emmitt, S., Socio-economic and environmental barriers for a holistic asset lifecycle approach to achieve circular economy: A pattern-matching method. *Technological Forecasting and Social Change*, 2021. 170: p. 120798.

[73] Jayarathna, C., Ryley, T., Caldera, S., Shooshtarian, S., Wong, P., Zaman, A. and Ruiz, A., Role of training in changing designers' attitudes and behaviour in optimal use of products with recycled content in the construction sector: A systematic literature review. *Resources, Conservation & Recycling Advances*, 2025. 25: p. 200248.

Chapter 6

Circular Economy Indicators for Construction Materials

Ana María Cáceres Ruiz and Atiq Zaman
School of Design and the Built Environment, Curtin University, Perth, Australia

Abstract

Companies aiming to deliver circular products to the market can transition to circular business models (CBMs), which are driving forces behind the adoption of circular economy principles. Companies can intentionally modify their value proposition to include circular strategies and adjust other key components of their business models, such as value creation, delivery, and capture. This chapter explores how the construction industry can embrace CBMs, to reduce resource depletion and waste generation, while setting the scene for a more circular-oriented industry. To do so, a literature review on CBMs and material circularity indicators was conducted. Next, a case study was used to assess how circularity can be measured and implemented in real-world settings. Findings on the circularity of construction materials (i.e., bricks) from an Australian company are provided and evaluated. Further recommendations on how to improve the circularity of these products are provided. In particular, the analysis highlights the potential for increased circularity through strategies like greater adoption of products with recycled content (PwRC), including those from construction and demolition (C&D) waste.

Keywords: circular economy, circular business models, material circularity indicator, construction industry, bricks

In: Policy, Practice and Education of the Circular Economy in Built Environments
Editors: Salman Shooshtarian, Peter SP Wong and Tayyab Maqsood
ISBN: 979-8-89530-623-9

1. Introduction

In a circular economy (CE) growth comes from within. Improving the value capture from existing products and materials will increasingly drive growth by providing multiple value-creation mechanisms decoupled from the continuous consumption of finite resources [1]. The CE paradigm promotes a shift from linear resource flows towards closed-loop resource flows that retain the environmental and economic value of products, thereby increasing resource efficiency and reducing raw material extraction and waste creation [2].

It is widely asserted by academia, government entities and societal stakeholders that business models play a pivotal role in the transition to a CE [3]. As the primary entities with the most prominent resources and capabilities [3], companies have the potential to significantly propel this transition if they embrace business models that not only make this move economically appealing [4] but also can compete with the linear take-make-dispose economy model [5]. Not implementing such Business models, will make the shift to a CE a fierce struggle [5]. In order to assist businesses in operating economically within a CE, the concept of a circular business model (CBM) has been developed [2]. Considering the substantial depletion of resources, waste generation, and ecological impact attributed to the construction sector, shifting towards the CE model has emerged as a paramount objective [6]. Some barriers to achieving CE models in this sector include resistance to change and the complexity of some of its products [6].

2. Research Objectives

This chapter aims to explore how Australian construction organisations can integrate CE principles into their business model/practice to foster the uptake of local products with recycled content (PwRC). Two main objectives were considered to achieve this aim:

1. Explore the circular business model practices that facilitate the uptake of local PwRC,
2. Identify drivers and enablers to foster the uptake of local PwRC in the Australian construction sector.

3. Research Approach

A comprehensive rapid literature review (RLR) approach was employed to conduct this research. RLR is a variant of systematic literature reviews (SLR) but stands out due to its significantly shorter timeframe for research execution (approximately 5 weeks), resulting in the rapid generation of credible evidence compared to SLR. The RLR methodology is endorsed by reputable organisations like the World Health Organization and is gaining momentum in various research domains, including building policy decisions to shape policy responses during the 2020 COVID-19 pandemic [7].

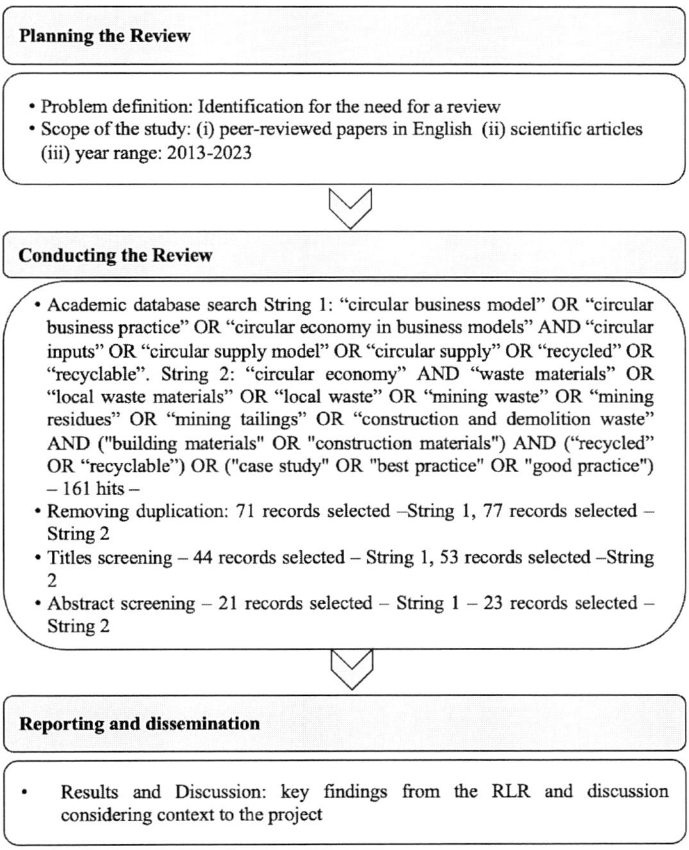

Source: Authors.

Figure 1. RLR process used in this research.

While RLRs are more commonly associated with the healthcare sector, their application has extended to other research domains, such as the field of building research [7]. One of the key strengths of this methodology lies in its ability to consolidate all available knowledge within a specific research area, facilitating a comprehensive understanding of the underlying theory. It transcends the mere creation of descriptive summaries of journal articles [8]. To mitigate bias and ensure a replicable review process, RLRs adhere to a well-defined and structured methodology [9]. This research, in its pursuit of rigour, embraced a three-stage approach to conduct a rapid review.

1. Planning the review
2. Conducting the review
3. Reporting and dissemination step [10], (see Figure 1).

Academic articles were collected from Scopus and Web of Science. Two search strings were used to tackle the research question (Figure 1). Literature from across the globe was considered during the search. Given that results from this platform are selected by relevance and number of citations, only the first three pages were taken into consideration. Other search criteria that were most relevant to the objectives included:

1. Only journal articles written in English were considered in the search criteria
2. Scientific articles included could be in the form of original research articles, reviews, conference proceedings, books, and book chapters
3. The search included articles from 2013 to 2023.

The articles resulting from the search strings were 161 (Figure 1). Duplicates were removed, and after the titles were verified for their relevance to the research inquiry, the abstracts were examined. If found suitable, a thorough analysis of the full papers was conducted. To enrich the pool of relevant studies, cross-referencing was applied within individual articles, allowing for the retrieval of additional background information related to specific research outcomes, concepts, or terms mentioned in the selected literature.

4. Circular Business Models

In CBMs new forms of value creation and capture emerge [11]. The value proposition in a CBM can be tailored to a product or service that intentionally implements a circular strategy to generate value [2]. A CE is underpinned by the 6Rs principles (often refers to a higher number of r-strategies), which mainly reduce, redesign, reuse, recycle, remanufacture, and repair processes of utilised products, by-products, and services. Supply chains can support and reinforce these principles; hence, they are vital in implementing CEs.

To understand how construction organisations in Australia can integrate circularity in their business model to foster local utilisation of PwRC, it is key to delve into the main elements of business models in the first place. A frequently used definition of business models is that of Osterwalder and Pigneur, which was adapted to better analyse CBM. This resulted in the 3-box framework. This model analyses CBM from the perspectives of value proposition, value creation/delivery, and value capture. Table 1 displays the questions required to address the value creation, value proposition, and value capture.

CBM accounts for an innovative avenue to realise economic growth by simultaneously enhancing resource efficiency and durability. CBM in the construction industry is still in its inception, mainly focused on meeting economic goals based on current operations and case-by-case project factors [6]. Significant changes are needed in the economic structure for adopting CE business models. These modifications should extend beyond individual projects and organisations, necessitating a well-defined circularity vision for engaging stakeholders in the procurement process [6]. Companies with a CE vision are committed to achieving sustainable long-term changes rather than pursuing short-term financial gains [12].

Table 1. Value dimensions and corresponding questions and elements

	Value creation & delivery	Value proposition	Value capture
Question	How is a value proposed?	What value is proposed and to whom	How does the company make a profit and capture other forms of value?
Key elements	Partners, resources, channels, activities	Customer segments/relationships Product/service offer	Cost structure Revenue streams

CBM strategies can be categorised into two main approaches: slowing and closing resource cycles [13]. While slowing resource loop approaches focus on the product use phase by prolonging the use and reuse of products, closing resource loop approaches are applicable between the after-use and production stages and aim to capture value from waste [13]. A CBM aims to maximise the use of products and materials and to extract their full value by keeping them in circulation for as long as possible. Two main circular business strategies exist: 'slowing resource loops' and 'closing resource loops.' The former focuses on product life extension. The latter includes extending the value of resources and promoting industrial symbiosis to harness the benefits of value propositions. The closing loops strategy emphasises significant reductions in greenhouse gas emissions achieved by adopting circular waste management practices. Implementing a CBM not only opens up new commercial opportunities whilst contributing to sustainable business growth but also shields the economy from resource shortages and rising material costs [14].

Circular business strategies can be explained through the three-box business model framework, including the (1) value proposition (product/service offering), (2) value creation and delivery (how value is provided), (3) value capture (how the firm makes money and capture other forms of value [15]. While introducing circular supply chain management initiatives holds great promise, these have not been implemented in the construction sector as they have been in other sectors (e.g., manufacturing) [14]. Circularity presents opportunities for businesses to drive sustainable development. Circular efforts should be evaluated based on their ability to contribute to the achievement of the United Nations' 2030 Agenda for Sustainable Development.

4.1. Drivers

There are potential solutions to better align CE principles with construction and demolition (C&D) waste management. The research community has widely suggested implementing standards for PwRC, advancing recycling technologies within enterprises, and implementing government interventions to promote the use of PwRC. Additionally, many countries have introduced new procurement models, such as public-private partnerships (PPPs), to address concerns related to C&D waste.

4.1.1. Public and Private Partnerships

Forming coalitions between the private and public sectors can also serve as an enabler of CBMs within the construction industry. Research has shown the interconnection between technical innovations and business model innovation in the construction industry. The shift towards circularity is slowed down mainly due to organisational constraints, not technical constraints. Governments and construction companies should be able to share risks in searching for new circular business opportunities. This collaborative approach can bring about both technical and business model innovations. For example, in cases where disruptive innovations don't align with existing tender requirements, collaborations between the private and public sectors can find alternative solutions to overcome such challenges [12].

4.1.2. Supply Chain

There is a strong link between CBMs and supply chains. The key distinction between traditional business models and those crafted for the CE is arguably most evident in their approach to creating and delivering value, with a specific focus on the supply chain [4]. Scholars are increasingly emphasising the need for a profound systemic transformation to achieve a CE, especially concerning the current supply chain structures [16]. In fact, supply chains are recognised as the fundamental operational unit for implementing the CE [14]. As a result, a new concept in sustainability has emerged, and it is known as Circular supply chain management (CSCM), which seamlessly integrates both forward and reverse supply chains, focusing on both closing and slowing resource loops. Some other CSCM definitions also include other resource loop management strategies such as narrowing, dematerialising and intensifying loops [4]. Adopting the CE in the supply chain yields various advantages, including reduced waste generation rates as well as improvements in resource efficiency, end-of-life management strategies, cost savings, organisational sustainability, and social benefits [14]. While introducing CSCM initiatives holds great promise, these have not been implemented in the construction sector as they have in other sectors (e.g., manufacturing) [14]. Companies have the potential to significantly propel the CE transition by generating additional value through an expanded and more strategically managed network of stakeholders within the supply chain [4]. Figure 2 below outlines the interplay between the 3-box BM framework, CBMs (slowing and closing loops) and the waste hierarchy (4R).

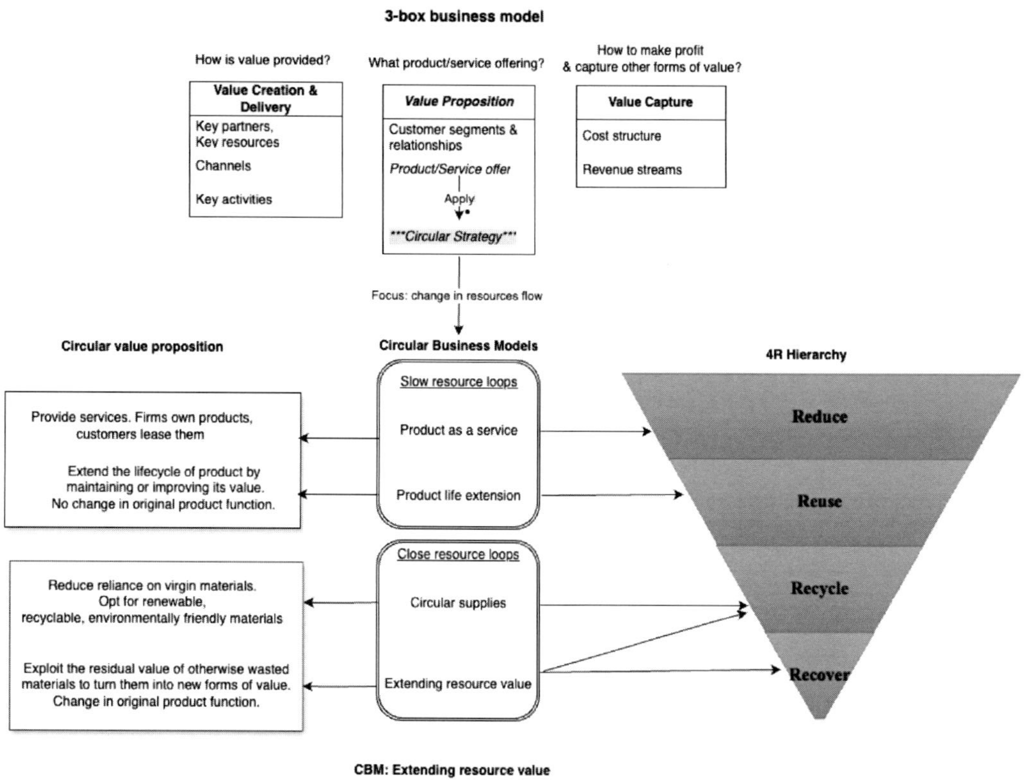

Adapted from Nußholz [2], Vermunt, Negro [11] and Bocken, De Pauw [13].

Figure 2. Interaction between the 3-box business model framework, main CBMs, and the 4R Hierarchy.

4.2. CBM: Use of PwRC

Value creation and delivery in the context of CBMs encompass a range of key activities. These include optimising procurement processes, implementing efficient separation at the source, organising the collection of waste materials by the material stream, and optimising sorting processes at material recovery facilities (MRF). Additionally, selective demolition practices are employed, and pre-demolition audits are conducted to identify reuse and recycling opportunities for materials. Take-back schemes are often utilised to efficiently collect and source materials for reuse. Key partners, such as collaborations with the market and supply chain, play a vital role in this process, emphasising the importance of public and private partnerships. To support these activities, key resources may be drawn from other companies, although the ideal scenario involves manufacturers developing business models that emphasise resource reuse (Figure 3).

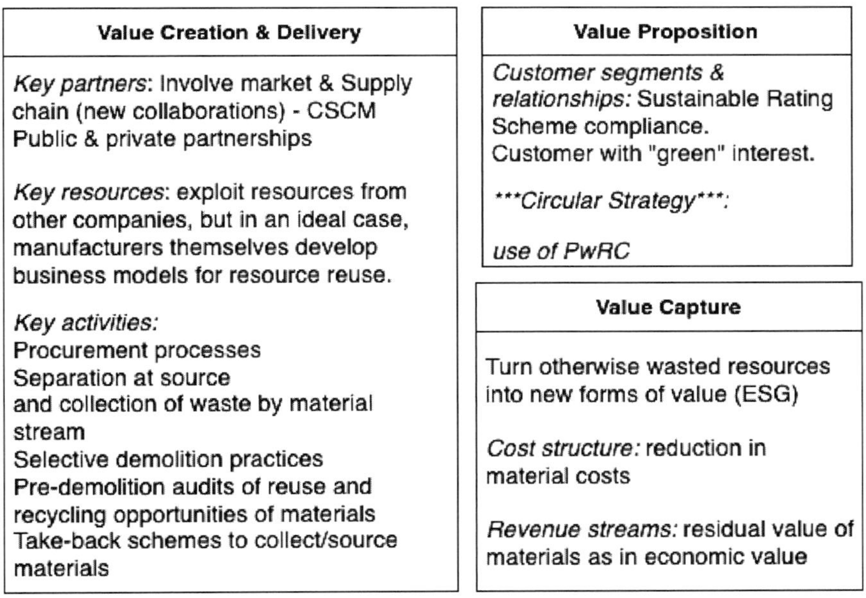

Sourc: Adapted from Bocken, De Pauw [13] and Vermunt, Negro [11].

Figure 3. CBM framework for circular value proposition "use of PwRC."

The value proposition of the business model revolves around achieving compliance with Sustainable Rating Schemes. This compliance serves as a significant factor in attracting specific customer segments and nurturing relationships with individuals who have a keen interest in environmentally sustainable practices and products. In terms of value capture, the business model focuses on transforming underutilised resources into fresh value opportunities. This approach reduces material expenses within the cost structure while simultaneously generating revenue streams from the remaining economic value of materials.

4.3. Challenges and Coping Mechanisms in Transition to a CBM

The key internal, external, market and institutional challenges and coping mechanisms in the transition to a CBM focused on extending resource value by offering PwRC are discussed below.

4.3.1. Internal Challenges

The internal barriers primarily revolve around knowledge and technology including limited awareness of how to use PwRC and the relevant recycling processes [11]. Additionally, there is a significant challenge in the absence of end-of-life (EoL) waste management initiatives supporting the integration of PwRC. To address these issues, organisations can focus on experimenting with advanced technology and knowledge development. Other solutions involve actively seeking knowledge from new sectors and, in certain cases, outsourcing technical activities to proficient partners [11].

4.3.2. External Challenges

External barriers primarily pertain to the supply chain. Organisations can face issues of overreliance on external parties for PwRC, leading to complications like uncertainty regarding quality, quantity, and delivery times of waste. The lack of partnerships and low waste volumes further compounded the challenges [11]. Coping strategies included building stronger, closer connections with supply chain partners to enhance product quality and reduce dependence on external entities. Another key approach was encouraging existing or potential suppliers to co-create circular products through collaboration and joint investment. Establishing credibility through research and raising awareness to reassure customers about product safety were essential aspects of managing these challenges.

Table 2. Barriers and strategies to implement CBM whose value proposition is focused on the use of PwRC

		Barriers	Coping/alleviation mechanisms
Internal	Knowledge & technology	Lack of knowledge on how to use PwRC and technology for recycling processes Lack of EoL waste management initiatives that facilitate the use of PwRC	Experimenting with technology and developing knowledge Searching for knowledge in new sectors Outsourcing technical activities
External	Supply Chain	Reliance on other parties for PwRC Uncertainty about quality, quantity, and delivery time of waste Lack of partnerships and low waste volumes Lack of awareness on the part of customers	Building strong, closer connections with supply chain partners to enhance product quality Maintaining control over product ownership to reduce reliance on external parties. Encouraging existing or future suppliers to co-create circular materials and products through collaboration and joint investment. Establishing credibility (demonstrate the additional value of PwRC by conducting research) Raising awareness (justify to people that products are safe)
	Market	Resistance from customers: distrust of PwRC Resistance from competitors Low prices of virgin materials	
	Institutional	Unsupportive waste legislations Waste is not a competitive resource: Lack of policy incentives to use waste.	Public-private partnerships (PPPs)

Source: Adapted from Mahpour [17] and Vermunt, Negro [11].

4.3.3. Market and Institutional Challenges

Market challenges can include customer resistance, often driven by distrust of PwRC, along with competitive resistance from industry peers. The prevalence of low prices for virgin materials further complicates market dynamics. Institutional barriers can also involve unsupportive waste regulations, particularly if waste is not regarded as a competitive resource, leading to a lack of policy incentives for waste utilisation. Organisations may explore PPPs as a means to address hurdles and promote circular practices, ensuring a smoother transition to a CE. Table 2 summarises the information presented in this section.

4.4. Key Findings of the CBMs

The points below describe the key findings of the CBM whose value proposition is focused on the use of PwRC:

- The CBMs are fundamentally different from the traditional business model as they focus on creating and capturing value. CBM strategies focus on both slowing and closing resource cycles.
- CBMs within the construction sector advocate for extending the use of materials and incorporating PwRC. Procurement processes assume a critical role in fulfilling these objectives, especially during the construction and demolition phase, because they actively promote the integration of PwRC in construction projects [6]. Embracing CBMs that make economic sense is essential for this transition.
- The concept of the CE emphasises improving value capture from existing products and materials, driving growth by decoupling it from the consumption of finite resources. It calls for a shift from linear resource flows to closed-loop flows and slow-loop flows to enhance resource efficiency and reduce raw material extraction and waste creation.
- Extending resource value is a strategy to close resource loops. Its value proposition focuses on exploiting the residual value of otherwise wasted materials to turn them into new forms of value. In addition, there is a change in the original product function. Using PwRC is a circular strategy that belongs to the "extending resource value" CBM.
- Implementing a CBM in the construction industry necessitates a significant shift in the current economic structure and requires a well-defined circularity vision.
- Circular supply chain management (CSCM) plays a crucial role in closing resource loops and achieving sustainability in resource use. Public and private partnerships can facilitate CBMs in the construction industry by driving innovation and overcoming organisational constraints.

Table 3. Material/Product-level circular indicators identified from the LR

Source	Abbreviation	Full form	Reference	Description
Academia	RPI	Reuse Potential Indicator	Park and Chertow [20]	Based on the concept of considering waste as a resource. Going from 'Waste' (RPI=0) to 'Waste-like' (RPI=0.4), to 'Resource-like' (RPI=0.75) and ultimately to 'Resource' (RPI=1).
	CEIP	Circular Economy Indicator Prototype	Cayzer, Griffiths [21]	The CEIP is initially intended to be used by manufacturing and/or retail companies of tangible goods with access to a bill of materials. CEIP assists in measuring the performance of their products against the EMF CE principles
	CI	Material-based Circularity Index	Cullen [22]	Accounts for the energy needed to recover a material, relative to the energy required to produce the material using virgin inputs, considering the recovered end-of-life material relative to total material demand.
	CPI	Circularity Potential Indicator	Saidani, Yannou [23]	Based on the MCI, CEIP and Circular Economy Toolkit (CET) from the University of Cambridge. Developed a CPI tool that through a guided questionnaire assesses the circularity potential of products covering phases such as design, development and benchmark
	PLCM	Product-Level Circularity Metric	Linder, Sarasini [24]	The approach is based on considering circularity as the fraction of a product that has been used before. Economic value is taken as the basic unit for aggregating product parts in a single value (PLCM)
	D^3R	Design for disassembly, deconstruction and reuse	O'Grady, Minunno [25]	Propose a method to assess the reusability of building components by calculating disassemblability, deconstructability and resilience of buildings and their components.

Table 3. (Continued)

Source	Abbreviation	Full form	Reference	Description
Academia	Sd	Level of Design for disassembly or the disassembly potential of a product	Jiang, Bhochhibhoya [18]	Adapt the MCI to tackle the indicator's two limitations. Firstly, the reliance on mass flow. Secondly, not considering the changes in the value of materials after end-of-life (residual value) - MCI assumes the value is the same as a new product.
	CCI	Circular Construction Indicator	Anastasiades, Blom [26]	The CCI framework helps assess a construction project throughout all construction phases namely design, construction, and end-of-life. Uses partial indicators (i.e., material scarcity, service life prediction) that can be adapted or replaced.
	M-CEFBC	Material Circularity Evaluation Framework for Building Construction	Amarasinghe, Hong [27]	The framework hinges on 15 sub-criteria allocated to 3 main criteria (Construction Management, Construction Design & Innovation, and Material Specification). The framework is a tool that helps in the selection of adequate construction materials, methods and technologies.
Legislative	EOL-RIR	End-of-Life Recycling Input Rate	European Commission [28]	The indicator measures how much of a material's input in the production system comes from the recycling of "end-of-life scrap" i.e., scrap and waste derived from the treatment of products at their end-of-life (EoL)
	MCI	Material Circularity Indicator	Australian Government (Department of Agriculture WatE) [29]	The Sustainable Procurement Guide of the Government of Australia advises on the use of the MCI to assess a product's circularity. MCI can assist in decision-making on material procurement and product design

Source	Abbreviation	Full form	Reference	Description
Business	NCI	New Circularity Index	EnelX [30]	The NCI developed by Enel X and ICMQ measures the significance of non-virgin materials and renewable energy consumed in a product's lifespan. It is calculated as a relative percentage of total material and energy flows. The NCI is applicable for cradle-to-grave scenarios. It requires data inputs such as the number of materials, water, energy and transports involved in the process, as well as the produced waste.
	CC	Circularity Calculator	IDEAL&CO Explore BV [31]	The circularity calculator was developed for designers working on circular products. The tool shows the flows of reuse, remanufacture and recycling, which facilitates visualisation.
	PC	Product Circularity	Cradle to Cradle Certified® [32]	This index is part of the Cradle to Cradle Certified® products program. The Material Reutilisation (MR) category was renamed Product Circularity. The MR index has been split into two sets of criteria: 1. Use cycled and/or renewable materials or communicate why this is not feasible, and 2. Use materials that are compatible with intended cycling pathways (i.e., recyclable, compostable, biodegradable).
Think tanks & advisory	MCI	MCI Material Circularity Indicator	Ellen MacArthur Foundation [33]	Measure the circularity of material flows for selected products. MCI supports decision-making on trade-offs between circularity and other company objectives on product design and materials procurement such as economic, environmental, and societal-related
	CTI	Circular Transition Indicators	The World Business Council for Sustainable Development [34]	CTI assessments commence with the completion of the full Close of the Loop segment, which measures how effective a company is at closing loops on its material flows. This is expressed as % material circularity, which is the weighted average of % of circular inflow (% non-virgin content and % renewable content (sustainably grown bio-based sources)) and % of circular outflow (% recovery potential (which is focused on design) and the actual recovery).

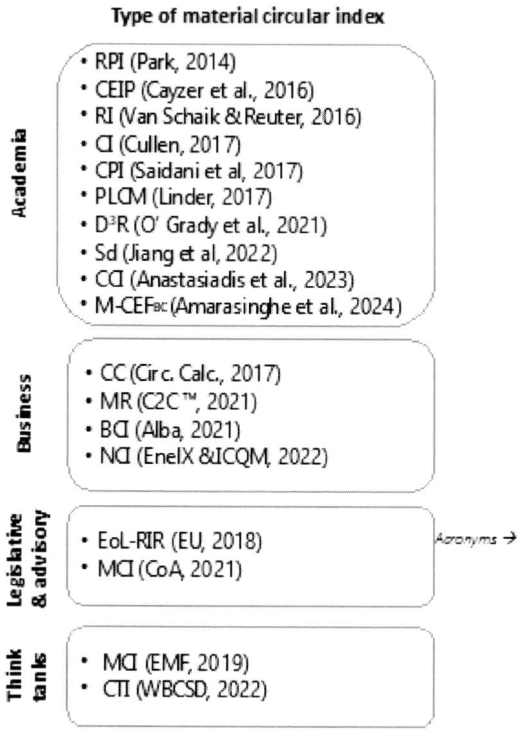

Assessment criteria of indexes (highest scored)

	Index	MCI	CEIP	CPI	NCI	CTI
Evaluation Criterion	Reduce	✔		✔	✔	✔
	RC	✔	✔	✔	✔	✔
	Recyclability	✔	✔	✔	✔	✔
	Renewables				✔	✔
	Recycling quality	✔		✔		
	DfD		✔	✔		✔
	Df Reuse	✔	✔		✔	✔
	Df Longevity	✔	✔	✔	✔	✔

*Other criteria may include local availability, design out of hazardous materials

RPI Reuse Potential Indicator
CEIP Circular Economy Indicator Prototype
RI Material-based Recycling Indices
CI Material-based Circularity Index
CPI Circularity Potential Indicator
PLCM Product-Level Circularity Metric
BCI Building Circularity Index
D3R Design for disassembly, deconstruction and reuse
CCI Circular Construction Indicator
M-CEFBC material circularity evaluation framework for building construction
CC Circularity calculator
MR Material Reutilisation
NCI New Circularity Index
EoL-RIR end-of-life recycling input rate
MCI Material Circularity Indicator
CTI Circular Transition Indicator
RC Recycled Content
DfD Design for Disassembly

Source: Authors.

Figure 3. Top clusters developing material circularity indexes cluster and circularity assessment.

4.5. Assessment of Material Circularity in Businesses

4.5.1. Overview of Material Circularity Indicators

Circular strategies, as delivered through new circular value propositions, should have a practical assessment approach in order to understand their impacts on the new CBM strategy. Measuring and identifying circular strategies have posed increasing methodological debates among scholars [18]. For example, Saidani, Yannou [19] found more than 50 circular indicators, which are differentiated according to implementation level, that is, macro, meso, or micro. There are many methods that practitioners, industry and think tanks have developed to assess product-level circularity [18]. As seen in the previous sections, there can be several strategies that can be the new circular value proposition of CBMs. Only by having adequate assessment frameworks that monitor progress towards circularity, can CE strategies be guaranteed [19]. Therefore, this section will answer the question of what are the most prominent material circularity indicators that would be most suitable to assess a real-life scenario through a case study analysis.

A rapid literature review was undertaken to gain a comprehensive understanding of material circularity indicators. Keywords such as 'material circular indicator (MCI),' 'material circularity index,' 'circular product index,' and 'material circularity assessment' were employed. The search was conducted using Scopus and Google Search databases. No temporal constraints were imposed, and only English-language literature was considered.

Table 4. An overview of highly scored MCIs

	Index	MCI	CEIP	CPI	NCI	CTI
Evaluation Criterion	Reduce	✔		✔	✔	✔
	Recycled content	✔	✔	✔	✔	✔
	Recyclability	✔	✔	✔	✔	✔
	Renewables				✔	✔
	Recycling quality	✔		✔		
	Design for disassembly		✔	✔		✔
	Design for Reuse	✔	✔		✔	✔
	Design for Longevity	✔	✔	✔	✔	✔

Source: Anastasiades, Blom [26] and Hallberg [35].

The selection process, which involved assessing abstracts and titles, yielded 24 journal articles selected for further analysis. Additionally, grey literature retrieved from Google Search was evaluated for relevance to the research question. A total of 9 articles and reports were deemed pertinent and included for subsequent review. The following sections outline the primary findings of the literature review and provide a rationale for recommending a specific material circularity indicator for evaluation in real-world settings supported by a case study.

4.5.2. The Rapid Review Findings of the MCIs

From the literature review on construction MCIs, it can be seen that this is a topic that has been studied mostly during the last decade, with increasing emphasis in recent years, particularly over the last 5 years. Research on MCIs comes from 4 main areas: - Academia, where the bulk of the research comes from; businesses; legislative and advisory institutions; and think tanks. Table 3 below provides an overview of the main material circularity indicators identified along with their description.

Some assessment criteria have been developed and proposed to ascertain the circularity of these indicators. The criteria can be derived from expert consultations, interviews, and extensive literature data. Among the criteria considered are recycled content, recyclability, design for disassembly (DfD), and design for reuse, among others listed in Table 4. This table identifies the criteria assessed by each indicator, with a particular focus on those achieving the highest scores. Notably, none of the indices comprehensively covers all the assessment criteria. It is important to underscore that these criteria are subject to constant refinement and expansion, particularly as they are heavily reliant on waste management processes. Further criteria can include economic and social considerations or the impact on regenerating ecosystems.

Table 5. MCI & LCA rough comparison

Metric	MCI	LCA
Focus	Material flow throughout use	Environmental impacts throughout the life cycle
Data input	Both require similar data input	
Complementary Impact indicators	The EMF recommends the use of additional indicators to understand the impact of increasing or decreasing MCI (e.g., LCA to assess energy consumption and CO_2 emissions)	
Future potential	Output parameter in LCA	Circular LCA: extend the boundaries of the traditional LCA and calculate the intended impacts for each successive life cycle of the product

Source: Ellen MacArthur Foundation [33].

4.5.3. The MCI: Recommended Circularity Product Tool

The MCI from the Ellen MacArthur Foundation (EMF) is widely used in academia and business settings. However, this indicator has some critiques, such as being very complex and data-intensive and not considering Design for disassembly, which has been identified as key when assessing the circularity of building materials.

While there have been many indicators proposed by different academics and other entities, the MCI from the EMF has been one of the most popular in the industry and academia - as seen in the rapid review -probably due to its simplicity and practicality. As it is user-friendly, even non-experts of CE can apply MCI [23]. Ellen MacArthur Foundation [36] has stated that there is no recognised path or approach to measuring how effective a product is in supporting the transition into circularity and developed the MCI for assessing product-level circularity [18]. The MCI is underpinned by lifecycle thinking principles; it accounts for the input, utility, and output. Considering that MCI does not demand extensive data, which happens to be a challenge when measuring circularity, MCI can provide a useful first overview of product circularity performance. On this note, MCI is a tool that can help industry practitioners make comparisons to determine the different material combinations [23], thereby assisting companies in their decision-making processes. These advantages have established the MCI as one of the most promising, practical [18] and popular frameworks for evaluating circularity performance. The MCI was updated in 2019 [33] attending to its high popularity.

4.5.4. MCI and Life-Cycle Assessment (LCA)

While LCA concerns environmental impacts, MCI focuses on encouraging the use of PwRC, as well as durability and increased utility, such as usage intensity, of a product [33]. These factors are not explicitly considered in LCA. Both LCA and MCI require region-specific assumptions and consider a product's lifecycle from creation to disposal. Table 5 below shows further analysis of both MCI and LCA.

The EMF calls for differentiating the design intent of a circular system and its actual performance. An item can be designed to be recycled, but the infrastructure needed may not be in place or products designed to increase durability may have banned substances [33]. Therefore, to make full use of the MCI, EMF recommends applying a multiple-life-cycle approach where the system is constantly monitored, and corrective actions are taken accordingly [33]. MCI could be further improved by including criteria that help identify the resource origin of biological material, account for the type of recycling (i.e., downcycling) and consider more recovery levels such as refurbishment [33]. Figure 3 provides an overview of the key findings from this section.

5. The MCI Assessment of Brick Manufacturing in Western Australia

5.1. The Case Study of BGC Bricks

BGC is a Perth-based company that manufactures construction materials and building products. BGC manufactures clay bricks in their two manufacturing facilities at (i) South Guildford and (ii) Middle Swan in Western Australia. The MCI analysis was conducted on two of their clay bricks product ranges: external bricks and internal bricks. Hereafter, the external bricks will be referred to as Face bricks and the internal bricks as Maxibricks (the main internal course brick). The sections below describe the input materials used in the production of these bricks and their current destination after use.

5.1.1. Feedstock Materials

The feedstock materials are categorised by virgin, reused and recycled input, as these are the key inputs to analyse with MCI using the EMF methodology [33]. The definitions of each of those inputs are as follows:

Table 6. Definitions of virgin, reuse and recycled inputs

Term	Definition	Reference
Virgin material	*"Material that is not from reuse or recycling."* It is also referred to as primary material.	[33, 37]
Reuse	*"The reintroduction of the same product for the same purpose and in its original form, following minimal maintenance and cosmetic cleaning. Reusing material implies avoiding the energy intensity and long supply chains of recycling" [33]. Additionally, diverting scrap (i.e., manufacturing scrap) from recycling is a reuse strategy.' [38].*	[33, 38]
Recycle	*"Recycling is the process of recovering materials to feed back into the process as crude feedstock. Recycling excludes energy recovery."* *"Recycling is the reprocessing of recovered materials at the end of product life, returning them into the supply chain. Recycled material may also be called 'secondary' in contrast to 'primary' material that is extracted from the environment." [37].*	[33, 37]

Therefore, it can be argued that:

> "Recycling is the reprocessing of recovered materials at the end-of-life to feed them back into the process as raw materials, returning them to the supply chain [33, 37]"

Virgin input: Clay: BGC uses around 20 types of clay from 15 different company-owned and leased clay pits.

Reused input: Batts: These are fired bricks or bricks that have undergone the kiln firing stage during brickmaking. These bricks don't continue in the process due to quality requirements. The fired batts are first crushed using a crushing line. This breaks them down into small particles, creating a sandy material. Then, they are fully blended into the clay mixes that are used in the Maxibricks production. The process described for the product Batts fits into the category of reuse for two primary reasons. Firstly, the process diverts the manufacturing scrap to an alternative rather than sending it for recycling. Secondly, although these bricks have been fired, they can be returned to the manufacturing process without affecting the structure and quality of the utility brick (Maxibrick).

Reclaimed clay: It is unfired clay classed as unsuitable for further processing due to quality requirements. Reclaimed clay is fed back into the start of the production process and blended with new clay. This is reused input because it is diverting manufacturing scrap from an alternate use to recycling and is being used in the same product.

Recycled input: Asphalt Granite Fines (AGF) is a by-product of asphalt production that comes in the form of rock dust. AGF does not require to be reprocessed before being fed into clay bricks, meaning that it keeps its original form, and it doesn't need to be treated for brickworks. Strictly adhering to the EMF definition of reuse, AGF is classed as recycled because is being utilised in a product for a different purpose, even if it is unprocessed.

The MCI includes the efficiency of the recycling rate (E_F) when there is recycled input. It follows the formula.

$$W_F = {M(1 - E_F)F_R}/{E_F}$$

For the purpose of this methodology, E_F is assumed to be ~99% because AGF is a by-product itself which is used unaltered in brickworks and has no prior reprocessing (or the like), generating no associated waste. So, W_F is assumed to be ~0.

5.1.2. The Output: The After-Use Phase

"The use phase of a product starts when it reaches its first users and ends when it is not reused any more as a whole. After the use phase, components can be reused and the rest of the product can go into recycling, energy recovery or landfill." [33]. There are three main material flows analysed in this phase: material collected for recycling, components collected for reuse and material going to landfill/energy recovery.

Material collected for recycling: In Perth, Western Australia (WA), Recycling of bricks is done through a mixed C&D waste stream, along with other materials such as concrete, wood and plastic. Very little to no bricks go to landfills. The vast majority are diverted to the hard waste recycling industry. C&D recyclers sort the mixed C&D waste to produce mono-material streams. The brick stream is then crushed and used to make recycled road bases and other aggregate products that meet specifications for uses like bulk fill, road bases, and landscaping. The highest quality recycled aggregate can be used to make a new road base. Lower-quality mixes may be used for other road base or bulk fill applications. Additionally, old clay pits are also often rehabilitated by filling them with crushed hard waste material rather than sending it to landfills.

In short, after-use bricks are crushed and recycled along with other demolition materials into new construction aggregate and fill products. While the material recovery rate for the C&D waste stream was 84 per cent in 2020–21, exceeding the 2025 waste strategy target of 77 per cent, this recovery rate is likely to be inflated due to the stockpiling [39]. To address this, this report focuses on the more specific masonry materials recycling rate, which was 76.4% for the same period [40]. This figure is deemed to reflect more accurately the number of bricks collected for recycling.

Regarding recycling efficiency rate, this research will consider the data from WA Recycling, a major C&D waste recycler in WA, with a reported rate of about 99% [41]. This figure is supported by other C&D waste recyclers in WA, such as Instant Waste, who separate the mix into clean, single-material streams which, according to reports provided monthly to BGC, go to uses other than landfilling.

Component collected for reuse: Currently, after the use phase, bricks are not being reused. Hence, this figure is assumed to be 0%.

Material going to landfill/energy recovery: This material corresponds to unrecoverable waste W_o, and can be calculated with $W_o = M(1 - C_R)$. While there is no evidence of bricks being used in energy recovery processes, a small number of bricks should be assumed to go to landfill (or left uncollected after demolition or improperly disposed of (i.e. illegally dumped), considering that they are handled in the C&D waste stream. The latest National Waste Report indicates that in WA for the year 2020-21, 518 tonnes of C&D waste went to landfill [42]. The MCI considers three after-use paths, recycling (in this case, 76.4%), reuse (in this case, 0%) and landfill, all of which should amount to 100%. Therefore, the percentage going to landfill is assumed to be 23.6%.

The Utility: The utility metric X comprises two components: product lifetime (L) and usage intensity (U). The former accounts for the length of the product's use phase (lifetime) and the latter for the intensity of use (functional units). The lifetime L/L_{av} assess the impact of the product's lifecycle on waste generation relative to the industry average L_{av}. The intensity of use component U/U_{av} reflects the extent to which a product is fully used compared to its potential.

$$X = \left(\frac{L}{L_{av}}\right) * \left(\frac{U}{U_{av}}\right)$$

It is expected that in most cases either lifetimes or functional units, but not both, will be used to calculate X. If lifetimes are used exclusively, U/U_{av} is 1. If functional units are used exclusively, L/L_{av} is 1 [33]. In this case study, bricks are considered a man-made rock that will last indefinitely, and this is no different from other bricks in the industry. Hence, U/U_{av} is considered 1 for the purpose of this calculation.

5.2. The Material Flows in BGC's Brick-Manufacturing

Figure 4 and Figure 5 depict material flows, which include the information above.

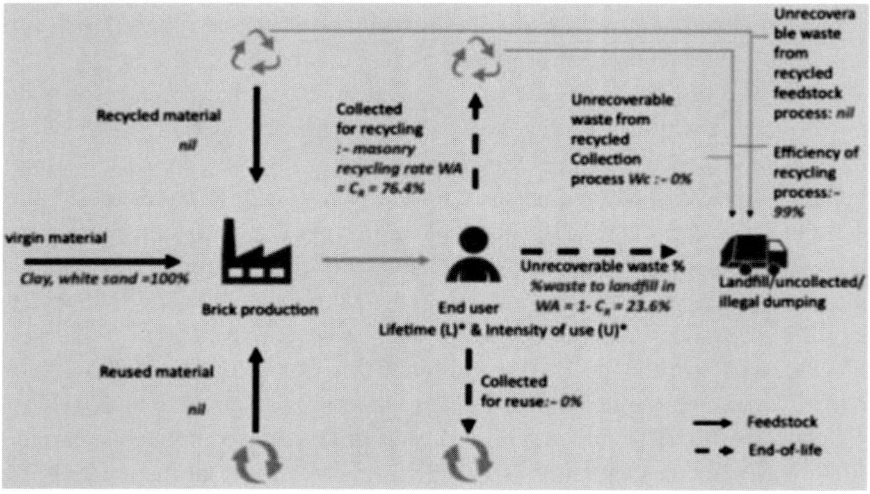

Source: Authors.

Figure 4. Technical flows of Face brick production in South Guildford and Middle Swan facilities.

Table 7 provides key data to determine the Face Brick and Maxibrick MCI baseline for both BGC South Guildford and Middle Swan facilities. Additionally, the MCI company level is seen in Figures 6 and 7.

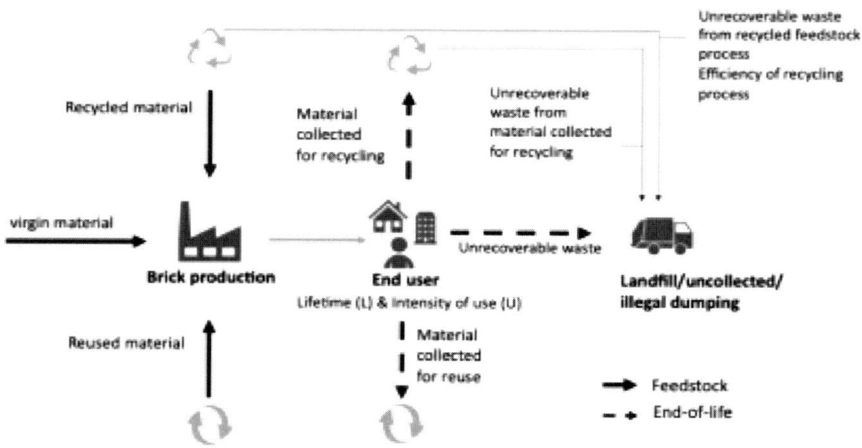

Source: Authors.

Figure 5. Technical flows of Maxibrick production in South Guildford and Middle Swan facilities.

Table 7. Face brick and Maxibrick characteristics - MCI baseline

	South Guildford		Middle Swan	
	Face Brick	Maxibrick	Face Brick	Maxibrick
Bill of materials	Clay Fine Sand	Clay AGF Reclaimed Clay Batts Fine Sand	Clay Fine Sand	Clay AGF Reclaimed Clay Batts Fine Sand
Feedstock materials	100% virgin materials 0% recycled materials 0% reused materials	86% virgin materials 1% recycled materials 13% reused materials	100% virgin materials 0% recycled materials 0% reused materials	93% virgin materials 1% recycled materials 7% reused materials
Destination after use	23.6% to landfill 76.4% to recycling 0% to reuse	23.6% to landfill 76.4% to recycling 0% to reuse	23.6% to landfill 76.4% to recycling 0% to reuse	23.6% to landfill 76.4% to recycling 0% to reuse
Baseline MCI	0.44	0.50	0.44	0.48

5.2.1. The MCI Baseline Scenario

The material specifications (virgin, recycled, or reused) and the end-of-life fate of the post-used bricks have been considered to measure the baseline MCI. Table 7 shows the parameters used to assess the MCI for both Facebrick and Maxibrick in BGC's two manufacturing facilities.

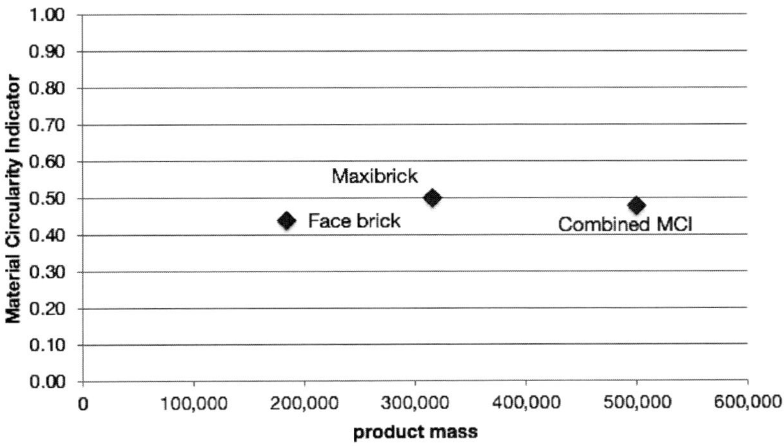

Source: Authors.

Figure 6. MCI per product and at South Guildford company level (Combined MCI = 0.48).

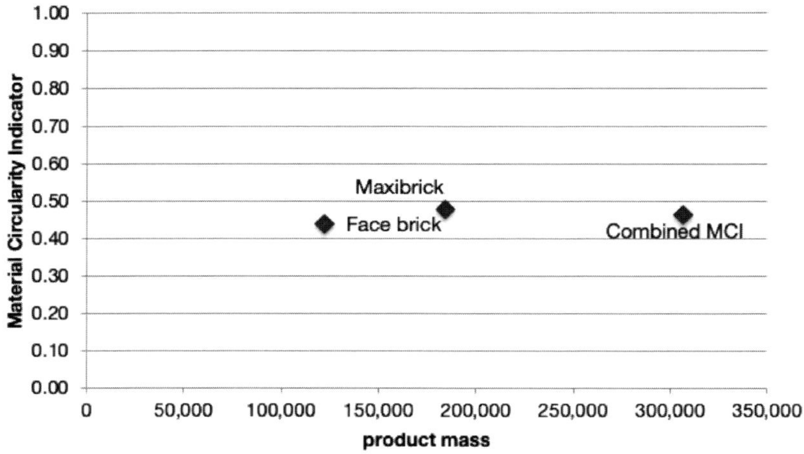

Source: Authors.

Figure 7. MCI per product and at Middle Swan company level (Combined MCI = 0.46).

The MCI of Maxibrick in the South Guildford plant shows a higher level of circularity performance, around 0.5, which is 50% of the circularity rate compared to the 44% of the Face brick (MCI = 0.44). Similarly, the MCI performances in the Middle Swan facility were measured as 48% and 44% for Maxibrick and Face Brick, respectively. The MCI performance shows (Figure 6 and Figure 7) a higher circularity performance of the South Guildford's brick (combined MCI = 0.48) compared to Middle Swan (combined MCI = 0.48).

The company-level MCI takes a reference product approach where each reference product represents a range of products with similar properties regarding material compositions, recycling and reuse rates, and utility (lifetime and functional units compared to an average product). For the case of the bricks produced in BGC, the Maxibrick represents one product range and the Facebrick, another. As there are no substantial differences with other types of bricks (i.e., Fastwall brick), in terms of composition, reuse, or recycling feedstock, these two types of bricks product range provide a good understanding of the MCI company level.

5.2.2. The Improvement Circularity Scenarios

The research also tries to hypothesise the improved circularity scenarios based on several assumptions, including:

- Increasing the recycling component to 10% for both facilities for Maxibricks
- Increasing reuse complements to 17% and 25% in the Middle Swan and South Guildford facilities, respectively for Maxibricks
- No changes in the recycling and reuse of Facebricks since the bricks use virgin materials to maintain their quality.

The MCI increases as the reuse and recycled feedstock increases. Using reused input benefits the MCI more than recycled input as it depends on the recycling efficiency of the process. The Facebrick is a product that requires different specifications than the Maxi brick particularly because is an external product. This could be challenging in the use of PwRC. Facebricks have much stricter quality requirements, especially around colour and appearance, than Maxibricks.To maintain these high-quality standards for the Facebricks, very little to no recycled materials in the BGC production process (like crushed fired batts or unfired reclaim clay) are incorporated into the Facebrick clay mixes. On the other hand, Maxibricks can incorporate a significant amount of these reused and recycled materials, as they have less stringent quality requirements. Below, some possible scenarios are considered to assess how the Maxibrick MCI could vary (Table 8). The improved scenarios of Maxibrick show a possible improved MCI of 0.6 from the current 0.5 and 0.56 from the current 0.44 for the South Guildford and Middle Swan facilities, respectively (Table 9).

Table 8. Maxibrick baseline and scenarios

	South Guildford			Middle Swan		
	Reused Input (%)	Recycled Input (%)	MCI	Reused Input (%)	Recycled input (%)	MCI
Baseline	13	1	0.50	7	1	0.48
Scenario 1	25	1	0.56	17	1	0.52
Scenario 2	25	5	0.57	17	10	0.56
Scenario 3	25	10	0.60	17	10	0.56

Table 9. Facebrick baseline

	South Guildford			Middle Swan		
	Reused Input (%)	Recycled Input (%)	MCI	Reused Input (%)	Recycled input (%)	MCI
Baseline	0	0	0.44	0	0	0.44

Table 10. Possible avenues to increase MCI of BCG bricks

Suggestion	Description
Dedicated skip bins from instant waste in construction and manufacturing GC areas for bricks	C&D waste materials could be collected for recycling. To achieve this, some key activities should be in place: pre-demolition audits to identify possible materials for reuse and/or recycling; application of selective demolition practices (e.g., carefully removing materials to preserve others); and separation at source and collection of waste by material stream (see Figure 8).
Exploring potential uses in new markets	There may be potential options and markets for reusing after-use bricks. These include heritage-looking bricks, commonly used in classical architectural structures, or WA Brickmatch, a program that collects and reuses second-hand bricks, pavers, and roof tiles [43].
Concrete, cement or brick waste-based bricks	Waste materials from C&D sites can be utilised in brickworks. They must be pulverised first and then mixed with sand grains of different sizes (see Figure 8).
Geopolymer bricks	Geopolymerisation is a process that uses industrial by-products (e.g., fly ash, slag, clay) as precursor materials to produce inorganic polymers that have notable technical characteristics (e.g., mechanical strength, longevity) [44, 45]. This method contributes to decreasing reliance on virgin materials, reducing waste and increasing diversion from landfills [45]. Unlike fired brickworks, Geopolymerisation [46]. To produce geopolymer bricks, precursor materials may include fly ash (waste from the fired coal industry), recycled brick waste or recycled concrete waste. Although fly ash is often contaminated with chemicals and can pose significant air and water pollution risks when used as a binding agent in brickworks, these environmental hazards are mitigated, and the brick's quality is higher [47]. Recycled concrete waste and brick waste can be obtained from the construction and demolition site. The segregated waste must be cleaned, and any contamination removed. Recycled concrete and brick waste are then crushed and pulverised (see Figure 8). Another geopolymer brick option is slag-based. "Slag is a high-volume by-product from the ores metal smelting process" [48]. Some other studies have also assessed the use of plastic waste along with precursor materials as a more sustainable alternative to clay bricks [44].

Table 10 summarises the possible pathways to increase Maxibrick and Face bricks MCI.

5.2.3. Alternative Feedstock Brick Considerations

An assessment of five different bricks -i.e., fired clay brick, fly ash cement brick, fly ash geopolymer brick, recycled concrete waste geopolymer brick, and recycled brick waste geopolymer brick- revealed that fired clay bricks contribute a higher environmental impact per compressive strength compared to alternative bricks. Using a multi-criteria decision-making methodology that considers technical, environmental, social and cultural aspects, the preferred choice is recycled waste bricks followed by fly ash cement bricks [45]. In Australia, although fly ash has a well-established use in cement production, less than half of it is recycled [49]. Geopolymer bricks are cost-effective, long-lasting and require less mortar during construction. They have high mechanical strength, use less energy to manufacture and recycle, are lighter, and absorb less water than other bricks (e.g., clay bricks) [47].

While the previous recommendations have shown promising paths to contributing to circularity in bricks, it is also key to assess financially, commercially, and socially any potential change to the present brick products. Most importantly, each alternative should be carefully tested and piloted before scaling up. Any approach should factor in market considerations, as they may face backlash compared to established practices. Collaboration amongst stakeholders, as well as quality assurance, are key to overcoming such resistance. Adjusting a product's MCI can bring about consequences that may affect a business and its stakeholders. Businesses can choose complementary indicators to determine the most relevant impacts [33]. While it is expected that increasing MCI results in greenhouse gas reduction, this should be assessed on a case-by-case basis. Well-established methodologies to do so include LCA and environmental product declaration (EPD) [33]. Other additional indicators may consider the changes in water use resulting from circular actions and potential toxicity exposure scenarios for circular products [33]. Economic considerations may include profitability assessment and identifying revenue and cost drivers [33].

5.2.4. Application of MCI in BGC's Products to Foster Circular Business

The baseline scenario of BGC's brick MCI performance helps understand the current circularity performance of the building materials. The baseline MCI would help understand their performance compared to industry practices/performance. It is evident that very few construction products available in the market inform their MCI. Thus, the MCI of the selected products may offer BGC competitive advantages over its competitors and pioneer its circularity strategies with specific targets and visionary improved circularity goals.

Furthermore, BGC should consider how to improve value creation and delivery, value proposition and value capture from the brick-manufacturing by incorporating circularity principles and strategies in BGC's business. Figure 8 shows the possible consideration of BGC's CE business model for brick manufacturing.

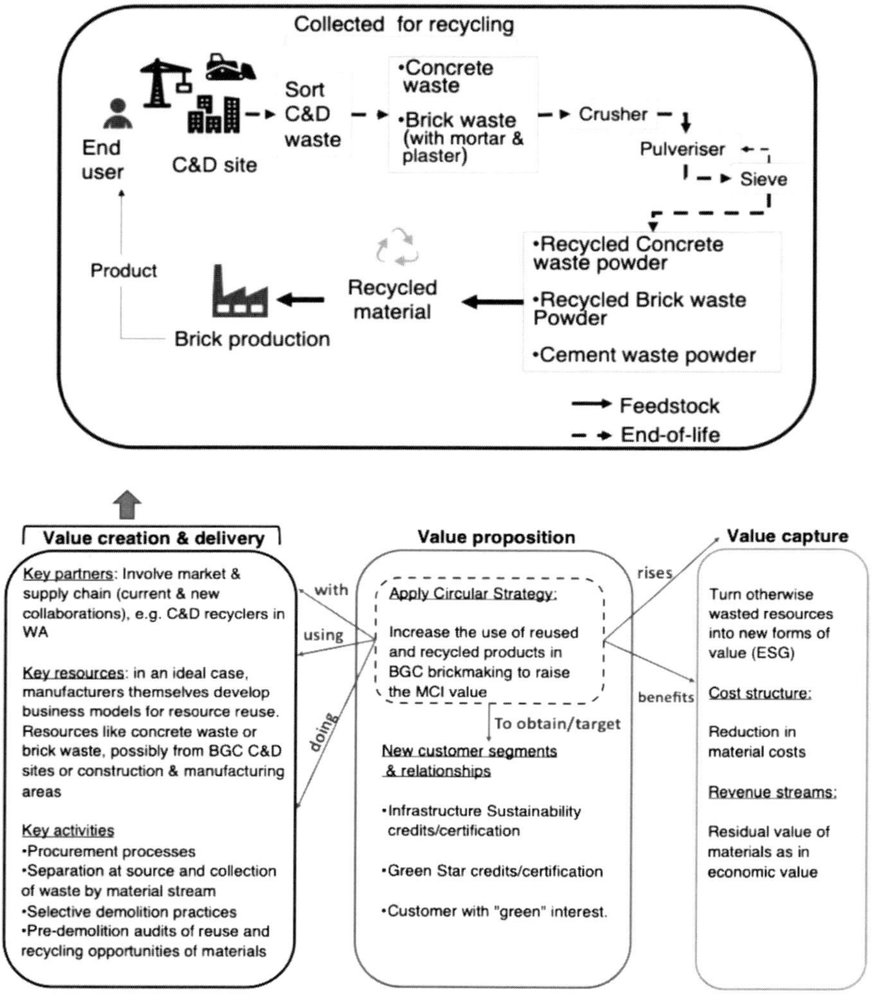

Figure 8. Top: Possible resources and avenue recovery to increase recycled content in BGC bricks, adapted from Maaze and Shrivastava [45]. Bottom: CBM strategy: Increase BGC bricks MCI by extending resource use (reuse, recycle).

The 3-box business model indicates how BGC can incorporate the circular strategy to increase their bricks' MCI. The strategy could boost value capture and benefit the cost structure. Additionally, the strategy could help obtain Infrastructure Sustainability and Green Star credits and certifications. Moreover, it could attract customers who value environmentally friendly products. The delivery of the strategy must consider key partners (e.g., WA

C&D recyclers), key resources (e.g., concrete or brick waste), and key activities (e.g., separation at the source, exploring new markets for reused bricks, such as heritage-looking bricks). These three essential pillars for creating and delivering this strategy can be supported by using resources available at C&D sites or construction and manufacturing areas. Figure 8 at the top exemplifies this process. Waste is sorted and processed, and then the resulting products - including recycled concrete waste powder, recycled brick waste powder, or cement waste powder - can all be used as feedstock in BGC brickworks, contributing to supporting the MCI.

5.2.5. Benchmark Comparison

This research marks the first attempt in Australia to measure and benchmark product-level material circularity, setting a crucial baseline for the country's brick industry. While it is challenging to make direct comparisons within the Australian context due to the lack of prior studies, parallels can be drawn with global standards. For example, Dräger, Letmathe [50] reported that the MCI for masonry brick and face brick in Germany was 0.52 and 0.47, respectively, demonstrating that BGC's bricks are performing on par with European benchmarks. This is significant for the Australian brick industry, as it underscores the potential for local manufacturers to align with or even exceed international standards. By integrating CE strategies into its business model, BGC not only strengthens its competitive edge but also positions itself as a potential leader in sustainable brick production. The findings offer valuable benchmarking opportunities for other Australian manufacturers, fostering innovation and driving the industry towards a more circular and sustainable future.

Conclusion

This research aimed to explore how Australian construction organisations can integrate CE principles into their business model/practice to foster the uptake of local PwRC. To do so, a literature review on CBMs was conducted. CBMs are a driving force behind the adoption of CE principles, particularly the focus on local PwRC. Companies can intentionally modify their value proposition to include circular strategies and accordingly modify other key components in the business models, such as value creation, delivery, and capture.

The construction industry, known for its resistance to change, faces unique challenges in embracing CBMs. The intricacy of construction products and limited digitalisation compound the difficulties of this transition. However, recognising the potential of CBMs to make this move economically appealing is essential, as they can compete with the conventional linear economy model.

This research underlines the potential of CBM to revolutionise the construction industry, reduce resource depletion and waste generation, and pave the way for a more circular-oriented industry. By addressing the challenges (e.g., distrust of PwRC) and leveraging the coping mechanisms identified (e.g., PPP), construction organisations in Australia can contribute to the transition to a CE while promoting the use of local PwRC.

This research analyses the material flows within BGC's brick production process to assess its circularity by calculating the MCI at the product level. The focus is on two primary brick types: Face bricks and Maxibricks. The feedstock is classified as virgin, reused, and recycled materials, and the end-of-life flows are categorised into those collected for recycling, collected for reuse, and unrecoverable waste. Overall, Maxibricks outperform Face bricks in terms of the MCI, with values of 0.5 and 0.44, respectively, at the South Guildford plant, and 0.48 and 0.44, respectively, at the Middle Swan plant. The MCI outputs provide groundbreaking insights into the circularity performance of BGC's brick products. For similar bricks, similar MCI performances were revealed in a European study. The research found that while some materials are reused within the production process (e.g., batts), the overall reliance on virgin materials is high. Additionally, the recycling rate for end-of-life bricks, while not low, could be much further improved by applying any of the suggestions within this report. The analysis highlights the potential for increased circularity through strategies like greater utilisation of recycled materials, including those from C&D waste.

To improve circularity, the research proposes exploring options such as using concrete waste, brick waste, fly ash cement or geopolymer bricks, as well as implementing dedicated waste collection systems for end-of-life concrete and bricks in construction and demolition sites. Additionally, further research focussing on the life cycle assessment of the baseline MCI and improved MCI scenarios could provide more details on environmental performance in relation to carbon emissions and other environmental parameters.

References

[1] Ellen MacArthur Foundation, *What is a circular economy?* . 2019.

[2] Nußholz, J. L., Circular business models: Defining a concept and framing an emerging research field. *Sustainability*, 2017. 9(10): p. 1810.

[3] Palmié, M., Boehm, J., Lekkas, C.-K., Parida, V., Wincent, J. and Gassmann, O., Circular business model implementation: Design choices, orchestration strategies, and transition pathways for resource-sharing solutions. *Journal of Cleaner Production*, 2021. 280: p. 124399.

[4] Geissdoerfer, M., Morioka, S. N., de Carvalho, M. M. and Evans, S., Business models and supply chains for the circular economy. *Journal of Cleaner Production*, 2018. 190: p. 712-721.

[5] Ranta, V., Aarikka-Stenroos, L. and Mäkinen, S. J., Creating value in the circular economy: A structured multiple-case analysis of business models. *Journal of Cleaner Production*, 2018. 201: p. 988-1000.

[6] Illankoon, C. and Vithanage, S. C., Closing the loop in the construction industry: A systematic literature review on the development of circular economy. *Journal of Building Engineering*, 2023. 76(2023): p. 107362.

[7] Eon, C., Breadsell, J. K., Byrne, J. and Morrison, G. M., The discrepancy between as-built and as-designed in energy efficient buildings: a rapid review. *Sustainability*, 2020. 12(16): p. 6372.

[8] Okoli, C. and Schabram, K., A guide to conducting a systematic literature review of information systems research. *Communications of the Association for Information Systems*, 2015. 37(43): p. 879-910.

[9] Grant, M. J. and Booth, A., A typology of reviews: an analysis of 14 review types and associated methodologies. *Health Information & Libraries Journal*, 2009. 26(2): p. 91-108.

[10] Higgins, J., Cochrane handbook for systematic reviews of interventions. *Cochrane Collaboration and John Wiley & Sons Ltd*, 2008.

[11] Vermunt, D. A., Negro, S. O., Verweij, P. A., Kuppens, D. V. and Hekkert, M. P., Exploring barriers to implementing different circular business models. *Journal of Cleaner Production*, 2019. 222: p. 891-902.

[12] Ruiter, H., De Feijter, F. and Wagensveld, K., Management control and business model innovation in the context of a circular economy in the Dutch construction industry. *Sustainability*, 2021. 14(1): p. 366.

[13] Bocken, N. M., De Pauw, I., Bakker, C. and Van Der Grinten, B., Product design and business model strategies for a circular economy. *Journal of Industrial and Production Engineering*, 2016. 33(5): p. 308-320.

[14] Lahane, S., Kant, R. and Shankar, R., Circular supply chain management: A state-of-art review and future opportunities. *Journal of Cleaner Production*, 2020. 258: p. 120859.

[15] Kohl, H., Seliger, G. and Dietrich, F., *Manufacturing Driving Circular Economy: Proceedings of the 18th Global Conference on Sustainable Manufacturing, October 5-7, 2022, Berlin*. 2023: Springer Nature.

[16] Kirchherr, J., Yang, N. H. N., Schulze-Spüntrup, F., Heerink, M. J. and Hartley, K., Conceptualizing the Circular Economy (Revisited): An Analysis of 221 Definitions. *Resources, Conservation and Recycling*, 2023. 194: p. 107001.

[17] Mahpour, A., Prioritizing barriers to adopt circular economy in construction and demolition waste management. *Resources, Conservation and Recycling*, 2018. 134: p. 216-227.

[18] Jiang, L., Bhochhibhoya, S., Slot, N. and de Graaf, R., Measuring product-level circularity performance: An economic value-based metric with the indicator of residual value. *Resources, Conservation and Recycling*, 2022. 186: p. 106541.

[19] Saidani, M., Yannou, B., Leroy, Y., Cluzel, F. and Kendall, A., A taxonomy of circular economy indicators. *Journal of Cleaner Production*, 2019. 207: p. 542-559.

[20] Park, J. Y. and Chertow, M. R., Establishing and testing the "reuse potential" indicator for managing wastes as resources. *Journal of Environmental Management*, 2014. 137: p. 45-53.

[21] Cayzer, S., Griffiths, P. and Beghetto, V., Design of indicators for measuring product performance in the circular economy. *International Journal of Sustainable Engineering*, 2017. 10(4-5): p. 289-298.

[22] Cullen, J. M., *Circular economy: theoretical benchmark or perpetual motion machine?* 2017, Wiley Online Library. p. 483-486.

[23] Saidani, M., Yannou, B., Leroy, Y. and Cluzel, F., How to assess product performance in the circular economy? Proposed requirements for the design of a circularity measurement framework. *Recycling*, 2017. 2(1): p. 6.

[24] Linder, M., Sarasini, S. and van Loon, P., A metric for quantifying product-level circularity. *Journal of Industrial Ecology*, 2017. 21(3): p. 545-558.

[25] O'Grady, T., Minunno, R., Chong, H. Y. and Morrison, G. M., Design for disassembly, deconstruction and resilience: A circular economy index for the built environment. *Resources, Conservation and Recycling*, 2021. 175: p. 105847.

[26] Anastasiades, K., Blom, J. and Audenaert, A., Circular construction indicator: assessing circularity in the design, construction, and end-of-life phase. *Recycling*, 2023. 8(2): p. 29.

[27] Amarasinghe, I., Hong, Y. and Stewart, R. A., Development of a material circularity evaluation framework for building construction projects. *Journal of Cleaner Production*, 2024. 436: p. 140562.

[28] European Commission, *Contribution of recycled materials to raw materials demand - end-of-life recycling input rates (EOL-RIR)*. 2018.

[29] Australian Government (Department of Agriculture WatE), *Sustainable procurement guide: A practical guide for commonwealth entities* 2020.

[30] EnelX, *Circular certification product assessment methodology and technical manual 2023*. 2023.

[31] IDEAL&CO Explore BV. *Circularity calculator*. 2024; Available from: https://www.circularitycalculator.nl.

[32] Cradle to Cradle Certified®. *Cradle to Cradle Certified® Product Standard Version 3.1 to Version 4.0 Comparison: Summary of Modifications and New Requirements*. 2021; Available from: https://bit.ly/43VYmXV.

[33] Ellen MacArthur Foundation. *Circularity Indicators - An Approach to Measuring Circularity*. 2019; Available from: bit.ly/3Gd2qZS.

[34] The World Business Council for Sustainable Development, *Circular transition indicators V4.0 metrics for business, by business*. 2023.

[35] Hallberg, A., *Development of a product circularity indicator suitable for the EPD-scheme of construction products: A review of product circularity indicators and their potential applicability in EPDs for the construction sector*. 2023.

[36] Ellen MacArthur Foundation, *Growth within: A circular economy vision for a competitive Europe*. 2015.

[37] Worrell, E. and Reuter, M. A., *Handbook of Recycling: State-of-the-art for Practitioners, Analysts, and Scientists*. 2014: Newnes.

[38] Allwood, J. M., *Squaring the circular economy: the role of recycling within a hierarchy of material management strategies*, in *Handbook of recycling*. 2014, Elsevier. p. 445-477.

[39] Waste Authority, *Waste Authority annual report 2021–22*. 2022.

[40] Circular Australia, *Australian circularity benchmarks*. 2024.

[41] WA Recycling. *Making Waste Work For A Sustainable Future*. 2020; Available from: https://www.warecycling.com.au/about-us/.

[42] Blue Environment, *National waste and resource recovery report 2024*, E. Department of Climate Change, the Environment and Water, Editor. 2025: Canberra, Australia.

[43] WA Brickmatch. *Recycling Program*. 2022; Available from: https://bit.ly/4jDLxGf.

[44] ul Haq, M. Z., et al. *Circular Economy Enabler: Enhancing High-Performance Bricks through Geopolymerization of Plastic Waste*. in *E3S Web of Conferences*. 2023. EDP Sciences.

[45] Maaze, M. R. and Shrivastava, S., Selection of eco-friendly alternative brick for sustainable development; a study on technical, economic, environmental and social feasibility. *Construction and Building Materials*, 2023. 408: p. 133808.

[46] Davidovits, J., Environmental implications and life cycle assessment LCA of geopolymers. *Proceedings of the Geopolymer Camp*, 2015.

[47] Murugesan, P., Partheeban, P., Manimuthu, S., Jegadeesan, V. and Christopher, C. G., Multi-criteria decision analysis for optimum selection of different construction bricks. *Journal of Building Engineering*, 2023. 71: p. 106440.

[48] CSIRO. *Dry slag granulation: producing valuable by-products from waste*. 2022; Available from: https://bit.ly/4irPLj3.

[49] Miatto, A., Emami, N., Goodwin, K., West, J., Taskhiri, M. S., Wiedmann, T. and Schandl, H., Australia's circular economy metrics and indicators. *Journal of Industrial Ecology*, 2024. 28(2): p. 216-231.

[50] Dräger, P., Letmathe, P., Reinhart, L. and Robineck, F., Measuring circularity: evaluation of the circularity of construction products using the ÖKOBAUDAT database. *Environmental Sciences Europe*, 2022. 34(1): p. 13.

Chapter 7

A Review of Circular Economy Policies

Salman Shooshtarian[1]
Peter SP Wong[1]
Moein Behzad[1]
and Chamari Jayarathna[2]
[1]School of Property, Construction and Project Management, RMIT University, Melbourne, Australia
[2]Cities Research Institute, Griffith University, Brisbane, Australia

Abstract

This chapter provides an overview of circular economy policies in the context of the Australian construction industry. Also, this chapter relies on the existing literature to extract the relevant policies. Based on the results of the literature review, the policies of the circular economy in Australia are divided into two groups including direct and indirect policies. The existing literature that has been studied incorporates journal articles and government and industry reports that are open to access on online websites. The findings of the literature review reflect that 17 policies have been proposed to enhance the application of products with recycled contents in the context of the Australian construction industry including 7 direct policies and 10 indirect policies. The results of this chapter can be helpful for C&D waste recyclers, construction companies, and authorities to apply effective policies to improve circularity in the construction industry.

Keywords: circular economy, construction industry, products with recycled contents, policies, Australia, construction and demolition waste

In: Policy, Practice and Education of the Circular Economy in Built Environments
Editors: Salman Shooshtarian, Peter SP Wong and Tayyab Maqsood
ISBN: 979-8-89530-623-9

1. Introduction

Over the past few decades, there has been considerable growth in the construction and demolition (C&D) waste stream because of significant urbanisation driven by a rising population, movement and migration. In the absence of a standardised definition for C&D waste, this waste stream is typically defined as materials taken from demolition and construction sites that are unwanted or discarded However, the new circular mindset prefers to recognise this stream's materials as resources rather than waste. The National Waste Report 2022 [3] states that, in 2020-21, Australia generated 29 million tonnes of C&D waste, an increase of 61% since 2006-2007. Comprising 44% of the total waste, this waste source remains the main generator of Australian waste. To respond to this issue, one solution is for Australia to become a circular economy (CE) that would foster a sustainable built environment [4].

To date, numerous definitions of CE have been proposed [5, 6]. The Ellen MacArthur Foundation explains the CE as "an industrial economy that is restorative or regenerative by intention," a definition widely acknowledged and adopted by government officials, practitioners, as well as academics [7]. The European Parliament offers a simpler definition, describing CE as a model of production and consumption founded on "sharing, leasing, reusing, repairing, refurbishing, and recycling existing materials and products for as long as possible," which lengthens the life cycles of products and materials [8]. Within the building and construction sector, a CE is an economic system that replaces the standard 'end-of-life' (Eol) philosophy [9]; the emphasis is on minimising, inventively repurposing, recycling and reclaiming materials from distribution or manufacturing and consumption activities to make use of those materials as long as possible in the cycle, thereby minimising the misuse of natural resources [5, 6, 9]. In the context of C&D waste, CE provides a means by which to limit waste's environmental and economic costs, including landfill maintenance, mining of virgin materials, landfill levies, transportation expenses and illegal dumping [10].

Moving from a linear economy to a CE warrants substantial changes to current production and consumption systems [11]. One of the major barriers to implementing CE objectives is the lack of a supportive policy framework [12]. Hence, one critical area for a society wishing to achieve a CE is to develop supportive policies that inspire circularity across the materials supply chains.

In Australia, Australia's National Waste Policy [13] identified avoiding waste, improved resource recovery, and increased use of PwRC as core principles of CE [14]. To implement this Policy, the National Waste Policy Action Plan [15] sets targets and actions including an 80% average resource recovery rate from all waste streams by 2030, and to significantly increase governments' and industry's use of PwRC [16]. Although no specific targets in relation to the construction sector are provided in the Action Plan, the increase in PwRC in the sector including determining their use in road construction, and development of associated national standards and specifications to improve demand for them in buildings and infrastructure are highlighted in the Policy [14].

Three types of policy instruments assist the sector in applying circular economy principles to the sector's projects: administrative, economic and informative. Licenses, bans, benchmarks and voluntary agreements between industry and government are examples of administrative or regulatory instruments, while economic instruments include fees, subsidies, taxes and other charges. Informative instruments comprise labelling, reporting obligations, certification initiatives and awareness-raising campaigns [11]. The literature [17] suggests that policy instruments could help to stimulate the demand for PwRC. The optimal use of PwRC keeps resources in the loop, which follows CE principles.

2. Research Objectives

To bridge the existing gap in CE literature regarding the role of policies in driving the transition, a comprehensive review study was conducted with the following objectives.

1. To identify the most relevant CE policies with direct and indirect impacts on the use of PwRC in construction projects
2. To explore how these policies can create a demand for PwRC.

3. Research Approach

This section explains the research approach applied in the chapter. In the first step, the research team carried out a comprehensive literature review to extract key CE policies. This review found the relevant CE policies that affect the application of PwRC in the construction industry directly or indirectly. The sources of the review incorporate journal articles, government, and industry reports that are open to access on online websites. The search between databases was conducted by care to only identify the literature that reflects the CE policies that have impacted the application of PwRC in the sector within the past few years. The analysis of extracted studies was carried out through a thematic analysis that contributed to the identification of major policies divided into policies with direct and indirect impacts.

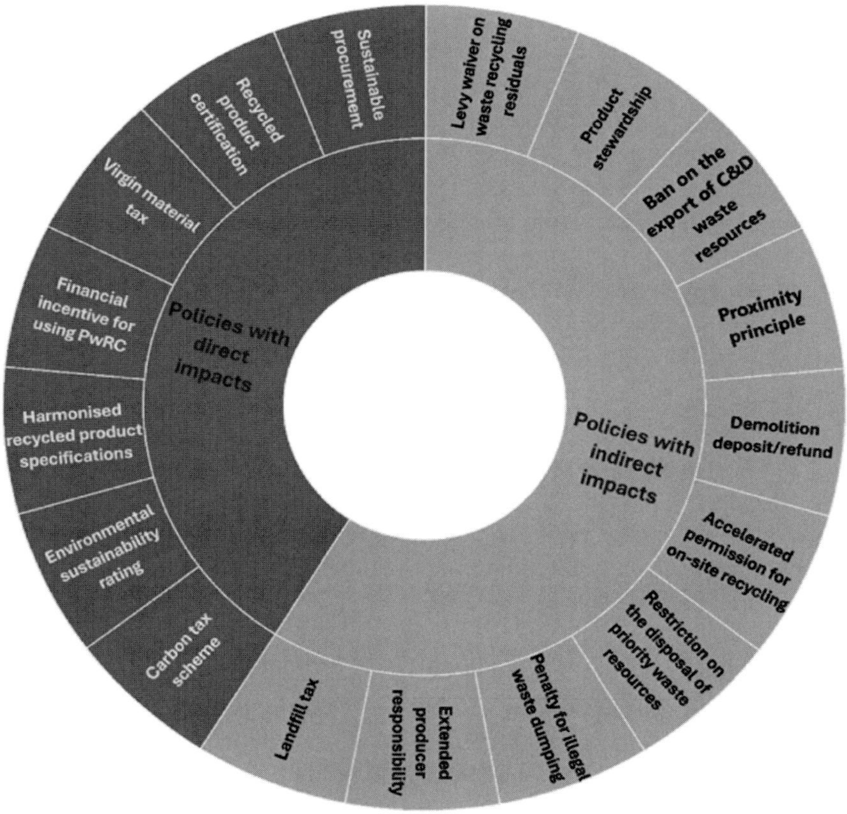

Source: Authors.

Figure 1. Circular economy polices (with direct and indirect impacts on the use of PwRC) investigated in this report.

Table 1. Summary of the study CE policies

Policy	Description	Policy	Description
Sustainable Procurement	Sourcing and purchasing recycled products while considering their environmental, social and economic implications	Penalty on Illegal Waste Dumping	A financial or legal consequence imposed on individuals, businesses or entities that engage in the improper disposal of waste materials in an unauthorised or environmentally harmful manner
Recycled Product Certification	A CE-based strategy that is designed to assure the quality, performance, environmental friendliness and safety of recycled products	Restriction on the Disposal of Priority Waste Resources	Regulatory measures and policies implemented by governments or authorities to limit or prohibit the disposal of specific types of waste materials that are deemed valuable, scarce, hazardous or environmentally damaging
Virgin Materials Tax	A financial mechanism that aims to discourage or penalise the use of virgin materials and encourage the use of recycled products	Accelerated Permission for Onsite Recycling	A regulatory or administrative process designed to expedite or fast-track the approval and permitting of recycling facilities or practices that are conducted on-site
Financial Incentive for Using PwRC	A monetary or economic benefit provided to individuals, businesses or organisations to encourage the use of recycled products	Demolition Deposit/Refund	Financial security or collateral mandated on eligible construction and/or demolition projects, with the primary objective of guaranteeing the appropriate recycling and disposal of C&D waste materials
Harmonised Recycled Product Specifications	Specifications designed to establish common guidelines and requirements that define the quality, composition and performance standards of products with recycled content	Proximity Principle	A waste management concept that emphasises the importance of handling waste as close to its source of generation as possible. It encourages minimising the transportation of waste over long distances and promoting the management of waste near its point of origin.

Table 1. (Continued)

Policy	Description	Policy	Description
Environmental Sustainability Rating (Recognition)	A systematic framework is used to assess, measure and rate the environmental sustainability of various entities, such as buildings, products, services and organisations. Examples include Green Star, LEED, and IS Rating.	Product Stewardship	A comprehensive and proactive approach to managing the environmental and social impacts of a product throughout its entire life cycle, from design and production to use and disposal
Carbon Pricing Scheme	It is an economic mechanism implemented by governments or regulatory bodies to incentivize and promote the use of recycled materials in construction projects while simultaneously imposing a tax or fee based on the carbon emissions associated with using virgin materials.	Ban on the Export of C&D Waste Resources	A regulatory policy or legal measure implemented by a government or authority to prohibit or restrict the export of C&D waste materials to other countries.
Landfill Levy (Tax)	A landfill levy is a financial charge or fee imposed by a government or regulatory bodies on the disposal of waste materials in landfills.	Recycling Residual Waste Levy Waiver	A waiver that allows a specific entity to be excused from certain requirements or obligations related to the disposal of residual waste materials when they can demonstrate a significant commitment to recycling practices
EPR	It is a policy approach that places the responsibility for the entire life cycle of a product on the manufacturer or producer, including the collection, recycling and proper disposal of the product once it becomes waste.		

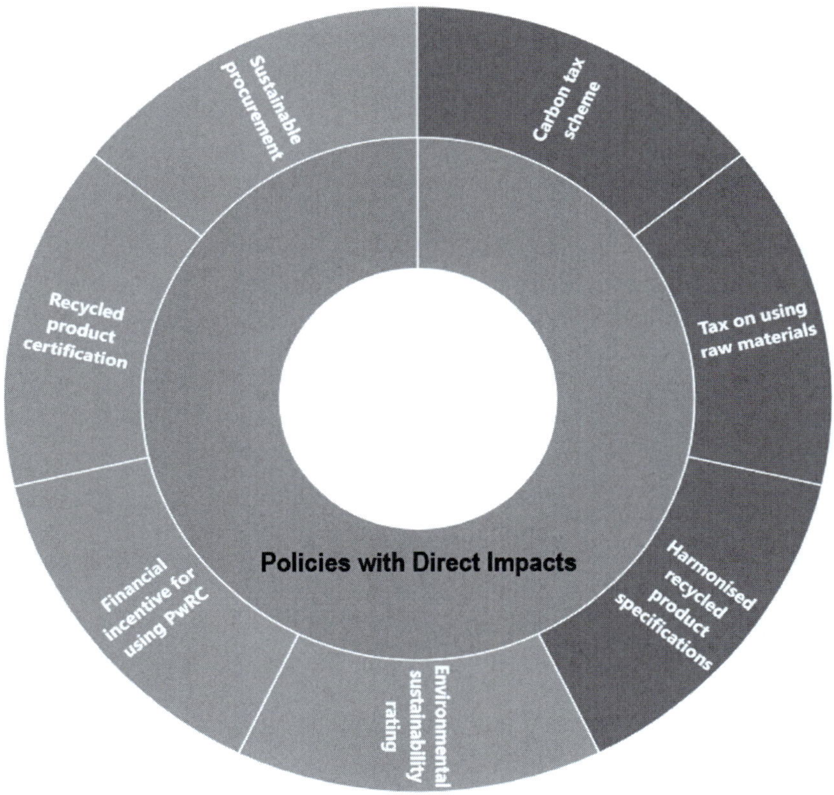

Source: Authors.

Figure 2. Policies with direct impacts on PwRC application.

3.1. Overview of Circular Economy Policies

In this section of the chapter, 17 policy interventions that are currently imposed or proposed to be implemented in the Australian context are explored and visually presented (Figure 1). These include 'levy tax on waste landfilling,' 'penalties on illegal waste dumping,' 'tax waiver on recycling residual,' 'tax on using raw materials' (economic), 'EPR,' 'sustainable procurement' (SP), 'proximity principle' (administrative), and 'recycled product certification' (RPC, informative). These policies are divided into two categories: policies with direct impact on the use of PwRC in construction projects and those that have indirect impacts. The subsequent sections provide an in-depth analysis of each policy. Table 1 presents a summary description of each policy.

4. Circular Economy Policies with Direct Impacts

In this section of the report, seven policies that have direct impacts on the procurement of PwRC in the building and construction sector are explored. Four of these policies are implemented in Australia, with varying degrees of application across states and territories. Figure 2 represents these policies.

4.1. Sustainable Procurement

Among current policies, sustainable procurement has the most potential to significantly improve C&D waste management systems. This policy approach will also help enhance circularity in resources in the building and construction sector. Various organisations define sustainable procurement for different scenarios and usages. The United Nations (UN), the UK government and the Australasian Procurement and Construction Council (APCC) accept the following definition: "*A process whereby organisations meet their needs for goods, services, works and utilities in a way that achieves value for money on a whole life basis in terms of generating benefits not only to the organisation but also to society and the economy, whilst minimising damage to the environment*" [18]. Sustainable procurement can also be referred to using terms such as 'green public procurement,' 'sustainable public procurement' [19], 'environmental purchasing,' 'circular procurement' and 'environmentally preferable procurement.'

Sustainable procurement adds to the concepts of CE and the cradle-to-cradle approach evaluated through various measures, including to evaluation product lifecycle. This approach is consistent with the principles of sustainable development, such as ensuring a strong, healthy and just society, living within environmental limits and promoting good governance' [20]. The existing research outlines many of the advantages of sustainable procurement applications. For example, Pick [21] lists advantages including energising the market at a local level, adding to the 'Zero Waste' movement and various goals relating to environmentalism and decreasing the costs faced by local governments using long-lasting and reusable materials.

In the building and construction sector, implementing sustainable procurement provides multiple environmental benefits. By way of example, it could lead to greenhouse gas (GHG) emissions reduction by curtailing energy use. This is important since estimations suggest that, of the energy needed to build a structure for a built environment, 80% is directed towards the production and transportation of construction materials [22]. Furthermore, sustainable procurement helps increase waste recovery and generates a demand for PwRC with a minimum environmental impact. A research study revealed that, among 25 study countries, mandating suppliers to commit to waste reduction goals is one of the five central functions of sustainable procurement policies [20]. A study in the Netherlands [23] reported that sustainable procurement is the primary approach to stimulating concrete recycling. Sustainable procurement was also the most promising option for establishing the long-term mechanism of using PwRC in the construction industry [24].

In recent years, only a handful of studies have investigated this topic; however, their scope was defined with some limitations. The limitations are concerned with the study focus or the methodology adopted. For instance, specific aspects such as carbon emission reductions [25], COVID-19 [26], target particular stakeholders [27] and market development [28] were previously studied. Using methodologies such as case study inherited constrained generalisability [29, 30] provides fractional information about building and construction sustainable procurement. The previous literature analysis shows that the relationship between procurement and C&D waste minimisation in Australia is not well studied [31]. Multiple research studies highlight the need to investigate the impact of procurement methods on waste management planning [32, 33]. Furthermore, the previous authors' exposure to real-time stakeholders' concerns expressed in their research projects has highlighted such shortcomings in the Australian building and construction sector's CE and waste management policy and practice. The stakeholders' concerns were captured in survey analysis [11, 34], a series of interviews [35], specialised workshops and industry seminars [36], and other personal communications. Considering all the above, this research was conducted to amend this research gap and offer a more comprehensive insight into this policy approach.

4.1.1. Sustainable Procurement Principles and Benefits in the Australian Context

To enhance the market demand for PwRC, government agencies need to become more involved in the wider implementation of sustainable procurement practices via the creation of specifications, quality assurance, accreditation and awareness raising regarding the outcomes (financial, social, environmental and ethical) of the PwRC and services [37]. The findings of a survey study [12] indicated that this policy approach, landfill levies and investment in technology and infrastructure are three major influential elements that significantly affect market development. In one study, research participants suggested that sustainable procurement is a vital solution for organisational waste management but only if it is a workable alternative or the client agrees to pay more [38].

In another study, Ershadi, Jefferies [27] explored the role of the project management office (PMO) in implementing sustainable procurement for construction projects. The study identified eight primary areas through which the benefits of this approach can be fulfilled. These include task assignment, strategic analysis, goal setting, tendering support, planning support, operation support, maintaining consistency and post-review. In the wake of the Australian government's 2010 Sustainable Procurement Guideline, a revised guide was released in 2020 [39]. This document outlines a framework for the Australian government to continue efforts to enhance sustainability outcomes and popularise sustainability principles in future procurement. This document offers a list of benefits to the purchaser (government), market, society and the environment that are fulfilled by the purchase of PwRC. Figure 3 [40, 41] integrates this list with the model proposed by the Australian ISO 20400 Committee.

Sustainable procurement principles are informed by 11 factors. As shown in Figure 4, these items refer to a range of facets of sustainable procurement and can guide public and private organisations on how to devise specific sustainable procurement policies.

Table 2 summarises the definition of these principles and proposes an action (the last column) that translates the principle into sustainable procurement of PwRC.

Table 2. Sustainable procurement principles, functionality and relevance to PwRC procurement

No	Principle	Description	Action relevant to PwRC procurement
1	Accountability	be accountable for its impacts on society, the economy and the environment, including the impacts of the organisation's supply chain	To increase the cost of utilisation of virgin materials for end-users to be accountable for their procurement policies and practices
2	Innovative solutions	seek solutions to address its sustainability objectives and encourage innovative procurement practices to promote more sustainable outcomes throughout the entire supply chain	To develop policies that promote innovation in the procurement of PwRC furthering PwRC uptake
3	Full and fair opportunity	avoid bias, and prejudice in all procurement decision making	To ensure that PwRC are not compared to virgin materials
4	Continual improvement	work towards continually improving its sustainability practices and outcomes, and encouraging its supply chains to do the same	To optimise the supply chain in the building and construction sector to embed PwRC in procurement planning and practices
5	Focus on needs	review demand, buy only what is needed and seek more sustainable alternatives first	To prioritise PwRC when specifying materials for low-value applications
6	Integration	ensure that sustainability is integrated into all existing procurement practices to maximise sustainable outcomes	To motivate the building and construction sector to build capacity for the application of sustainable procurement
7	Ethical behaviour	behave ethically and promote ethical behaviour throughout its supply chains	To monitor waste handling and processing activities to minimise the risk of procurement of PwRC.
8	Respect for stakeholder interests	respect, consider and respond to the interests of stakeholders impacted by its procurement activities	To provide a level playing field for all stakeholders involved in the use of PwRC
9	Transparency	be transparent about its procurement processes and how its decisions and activities impact the environment, society and the economy	To ensure transparency in handling, processing and procurement of PwRC
10	Life cycle costing	consider the cost incurred, the value for money achieved and also the costs and benefits on society, the environment and the economy resulting from its procurement activities	To consider the social and environmental benefits of procuring PwRC
11	Respect for human rights, the rule of law and international norms of behaviour	be aware of any violations throughout its supply chains and actively encourage its suppliers to do the same	To reduce the need for virgin materials through procurement of PwRC

Source: Local Government NSW (2021) [44].

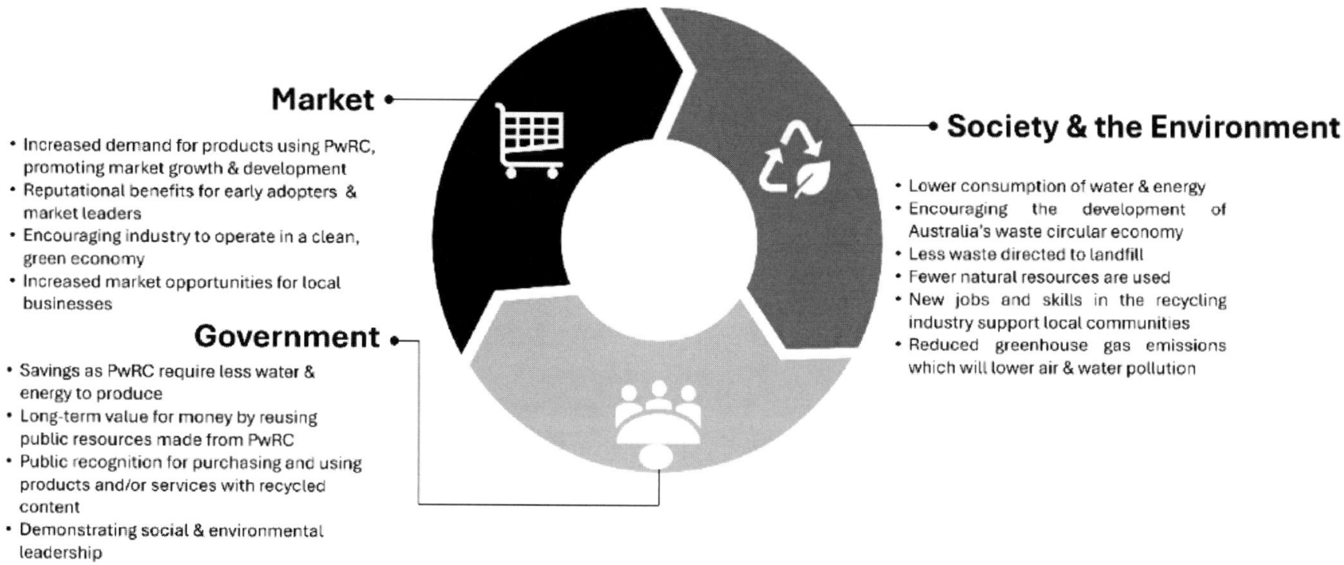

Source: adapted from [42, 43].

Figure 3. Advantages of sustainable procurement by society, government, market and environment.

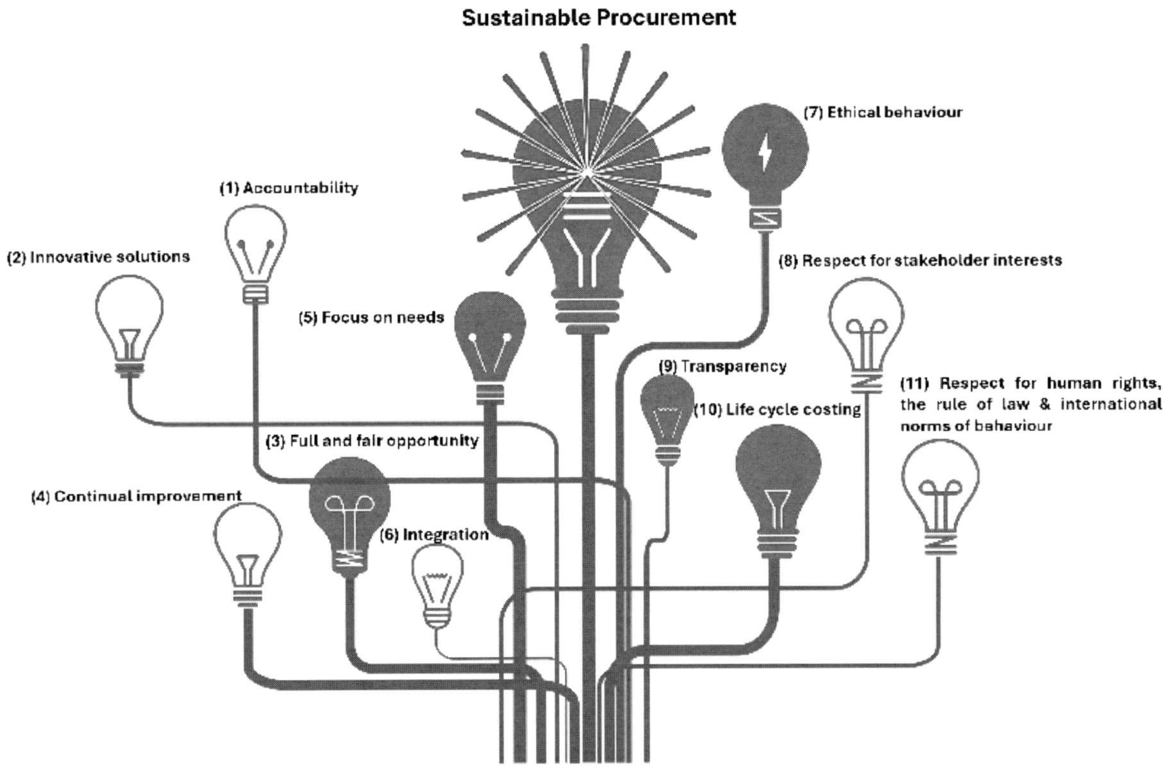

Source: Adapted from Local Government NSW (2021) [44].

Figure 4. Primary principles of sustainable procurement.

Table 3. Summary of barriers to applying sustainable procurement building and construction sector

Ref.	Study context	1	2	3	4	5	6	7	8	9	10	11	12	13	14	15	16	17	18	19	20	21	22	23	24	25	26	27	28	29	30
		lack of long-term planning	higher costs of PwRC	lack of methods to measure sustainability	lack of knowledge about sustainable products	obligation to procure from local suppliers to lower	unavailability of certain PwRC	lack of government support	misconception about PwRC production cost & time	lack of motivation	lack of supportive organisational culture	lack of inclusion of PwRC in construction specification documents	lack of mandatory imposition	uncertainty about PwRC quality	lack of operations	lack of an organisational policy	struggle to procurement behavioural change	lack of market & market barriers	traditional procurement procedures	oversupply of PwRC	lack of sustainable suppliers	time pressure to complete projects, minimising the possibility of sustainable procurement	difficulty in assessing the non-environmental impact	the complexity of cost-benefit analysis for PwRC	lack of training and education for suppliers	involvement of various stakeholders	shortage of relevant resources	suppliers focus on price rather than quality	desire to diversify contractors	economic uncertainty	lack of aptitude for sustainable procurement by top
[45]	Brazil	◆	◆		◆	◆		◆			◆														◆						
[46]	Brazil									◆	◆				◆			◆												◆	
[21]	Canada											◆		◆						◆		◆				◆		◆	◆		
[47]	France									◆	◆			◆	◆																
[48]	Malaysia	◆			◆			◆																			◆				
[49]	Saudi Arabia										◆																				◆
[50]	South Korea					◆	◆							◆																	
[41]	UN		◆										◆			◆		◆	◆												
[51]	Global								◆								◆				◆		◆	◆							

Table 4. Summary of enablers to apply sustainable procurement in the building and construction sector

Ref.	Study context	1 developing clear & supportive regulations	2 Project Owners' commitment to sustainability	3 project team capability	4 strategic partnership	5 customer expectation/pressure	6 providing incentives and rewards	7 strong technical expertise	8 quality assurance to maintain consistency	9 corporate reputation	10 proper planning & anticipating the arrangement	11 improving sustainability knowledge through education	12 setting a measurable target for using PwRC	13 maintaining transparency & good governance	14 mandatory application of sustainable procurement	15 conducting a cost-benefit analysis to educate project leaders & investors	16 potential to export locally manufactured goods	17 screening out suppliers with a lack of commitment to sustainability during the tendering stage	18 knowledge sharing among project stakeholders	19 effective coordination to address challenges and conflicts	20 procurement post-review
[52]	Australia								◆		◆		◆	◆		◆		◆		◆	◆
[27]	Australia																				
[21]	Canada		◆											◆							
[53]	China	◆				◆	◆														
[47]	France												◆								
[54]	Malaysia	◆	◆		◆		◆					◆			◆				◆		
[55]	New Zealand					◆				◆											
[56]	Nigeria	◆										◆		◆							
[57]	South Africa	◆			◆			◆									◆				

Source: Authors.

4.1.2. Barriers to Applying Sustainable Procurement

Barriers play a significant part in shaping sustainable procurement practices and determining the success or otherwise of sustainable procurement outcomes. The literature uncovers various elements that hamper the publicity and implementation of sustainable procurement practices. The challenges involved depend greatly on the specific scenario, though there are similarities between various contexts. Table 3 outlines studies that explore sustainable procurement barriers in developed and developing nations and presents 30 barriers reported in the literature. The two key barriers are revealed to be 'the lack of supportive organisational culture' and 'uncertainty about the PwRC quality.'

4.1.3. Enablers to Apply Sustainable Procurement

Procurement experts contend that sustainable procurement planning must be founded on the realisation of several mostly context-specific conditions. Previous studies (Table 4) have identified a range of enablers that would enable the effective implementation of sustainable procurement, mainly in the public sector in developed (i.e., Australia, New Zealand, Canada and France) and developing (i.e., South Africa, Saudia Arabia and Nigeria) nations (Table 4). From the 20 enablers identified in the literature analysis, the two key enablers were revealed to be 'developing clear & supportive regulations' and 'maintaining transparency & good governance.'

4.2. Recycled Product Certification

Recycled product certification (RPC) offers verifiable data on a PwRC's quality, safety, performance and environmental performance. A PwRC awarded an RPC after stringent evaluations and quality control can be perceived more positively, potentially attaining higher market value, uptake and buyer's confidence. As shown in Figure 5, key stakeholders in the building and construction sector benefit from the application of RPC. Recycling companies in the construction materials supply chain can gain advantages by using RPC to showcase product features required by potential buyers [58]. For builders, RPC serves as an efficient information management method to streamline workload in tenders requiring PwRC. Utilising RPC will allow builders to manage any risk that may arise from the usage of PwRC. In this circumstance, clients and contractors will bear the risk equally. Such schemes also enable sustainable procurement policies in private and public sectors domestically and develop the overseas market for states that recognise certain certificates. For clients, it offers an option for those who are willing to work towards sustainable development [59]. For enter the market in two forms: non-certified and certified. Most recycling plants produce non-certified PwRC. However, in recent years certification has become essential as consumers increasingly demand PwRC with specified and assured quality. It is important to understand that product standards and eco-labels are distinct. Eco-labels are supposed to tell consumers about the environmental impact of their purchases. Most eco-labels do not give consumers real information about the product, but rather they suggest that it is among the top 10-20% of its product category from an environmental standpoint [58].

It is highlighted that classification of PwRC and aggregates into easily comprehensible categories, together with correct certification, facilitates future client purchases [60], as they will be able to acquire a product suited to its intended use (e.g., structural concrete may use a recycled aggregate of class A, whereas recycled aggregate of class D may be used in road construction for subgrade). Currently, there are two categories of marketable PwRC: non-certified and certified. Non-certified PwRC constitute most of the recycling plants' output. However, certification is crucial due to consumers' increasingly stringent needs for PwRC of a specified and assured quality. It is important to understand that product standards and eco-labels are distinct. Eco-labels are supposed to tell consumers about the environmental impact of their purchases. Most eco-labels do not give consumers real information about the product, but rather they suggest that it is among the top 10-20% of its product category from an environmental standpoint [58].

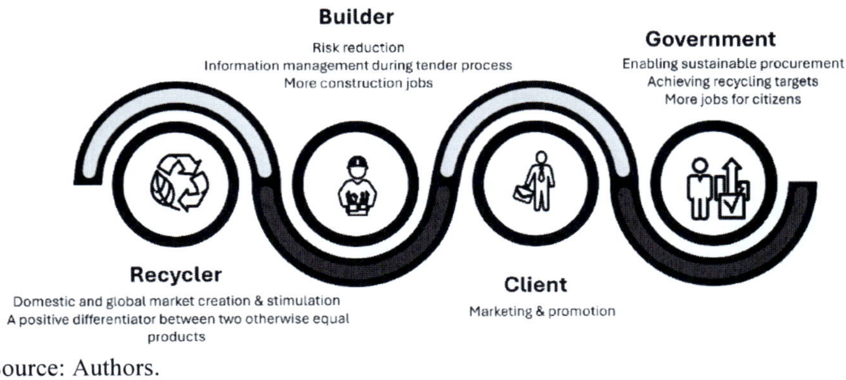

Source: Authors.

Figure 5. The advantages of RPC for recycled products supply chain participants.

4.2.1. Main Barriers to the Application of Product Certification

Despite the long history of product certifications [58], their full application in the building and construction sector is rather recent. Few studies detail the challenges associated with their practical use. The existing literature suggests that the implementation of the PRC is hindered by several challenges. Figure 6 shows that these challenges include high costs, currently unjustified expenses, reliance on self-certification, misalignment between RPC and sustainability rating systems, and the presence of unreliable certifiers.

A report by Equilibrium [61] indicated that in Australia, testing of PwRC for quality, performance and safety is inadequate, resulting in these products lacking independent validation. Further, the scarce opportunities to use these products are inadequately advertised, making it hard for stakeholders to justify the expense of certifying such products. Ghaffar [62] indicated that mandatory RPC expenses in the EU add to the price of the material and may negate any cost savings from reusing it. Self-certification is commonly used to establish PwRC quality, with recyclers and PwRC suppliers often performing it themselves. While more cost-effective, this approach lacks the assurance that end-users can safely rely on when procuring PwRC.

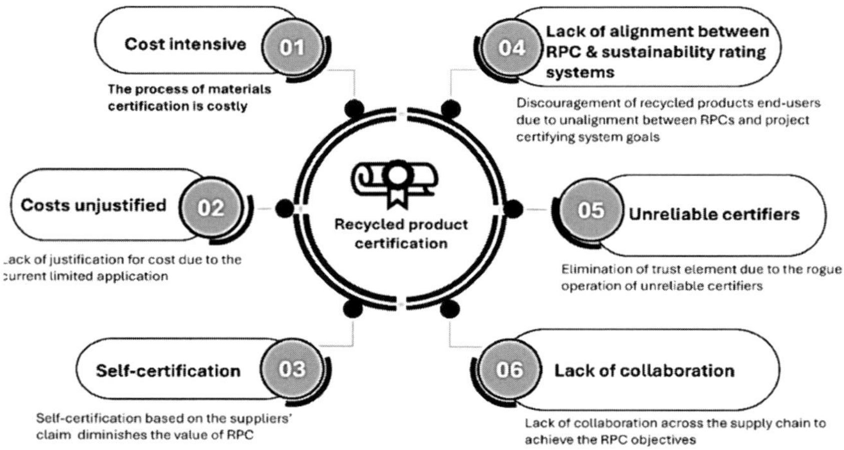

Source: Authors.

Figure 6. Main barriers to the application of product certification for PwRC.

4.2.2. Main Enablers of the Application of Product Certification

Six categories of RPC enablers are identified following a comprehensive literature review. These enablers are illustrated in Figure 7. Effective quality monitoring and certification of PwRC by suppliers are required to instil and maintain stakeholder trust in these products. However, this must be accompanied by more vigorous government action in the form of legislation and standardisation [59]. The government has a part to play in changing the general public's perceptions towards PwRC via promotional activities through the media. Ongoing training offered by the government could contribute to shifting the attitudes of industry stakeholders [63]. Furthermore, contractual obligation or another client requirement to use certified PwRC is another enabler for the implementation of RPC [58]. Currently, in the EU, it is illegal to use recycled content in construction materials that are not certified [61], which has resulted in more frequent uptake of these programs in construction projects across the European territory. According to a report by WSP Environmental and TRL Ltd [59], the benefits of mandatory RPC include the elimination of poor actors, improved stakeholder perception of data reliability, reduced opportunities for manipulation and enhanced credibility, all of which strengthen the value of RPC.

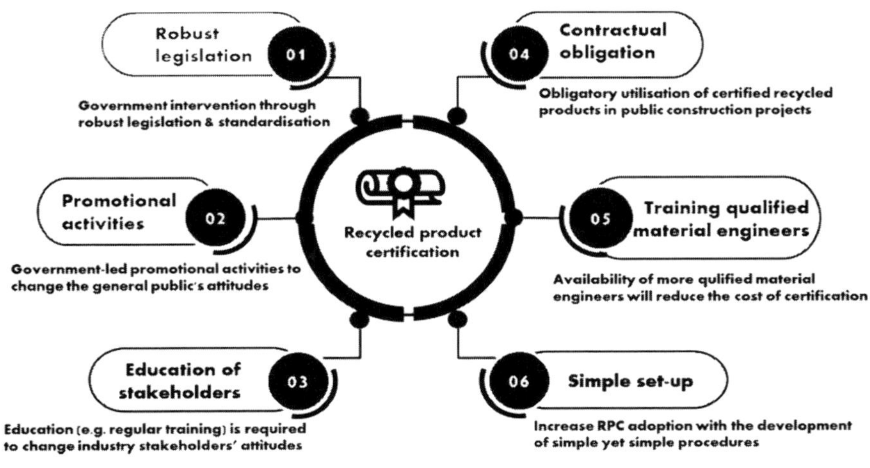

Source: Author.

Figure 7. Main enablers of the application of product certification for PwRC.

The fifth category of enablers refers to the availability of qualified material engineers at reasonable prices. According to anecdotal evidence, these professionals are scarce and expensive to employ. By educating more individuals to become trained materials inspectors, not only will jobs be created, but audit costs will also be reduced. The final category addresses the simplicity of certification methods for PwRC. The waste recovery industry's interest in RPC will be piqued by certification processes that are simple yet effective and require a reasonable amount of administrative work [58].

4.3. Tax on Using Virgin Materials

Although most construction materials are technically recyclable, the intention to recycle them and the extent to which recycling occurs often depend on their value. In most circumstances, this is compared to the price of a material when newly extracted or imported. To shift perceptions towards sustainably procuring PwRC, certain pricing mechanisms can be implemented to discourage buyers from selecting virgin materials for construction projects [64]. This enticement takes two forms— 'removing subsidies for virgin materials' and 'taxing on the use of virgin materials.' The application of such mechanisms in several countries has demonstrated improved competitiveness of PwRC in the construction materials market. For example, since 2002, the UK has implemented a regulation [65] to strengthen the position of PwRC relative to new materials. This regulation takes the form of a levy (£2 per tonne) applied to the use of gravel, sand and rock for commercial purposes. The purpose is to adjust the cost of new materials to reflect their underlying environmental impact. Sweden, Denmark and France, among other EU countries, have applied a similar tax [24].

In the Australian context, waste definitions that exclude clean fill have been adopted by a few states. Coupled with reduced costs, this approach is anticipated to encourage the building and construction sector to adopt PwRC rather than raw materials. Survey findings in Australia [34], suggest that key stakeholders perceive cheaper virgin materials as one of the five top barriers towards effective C&D waste management. Recently in the European Union, a tax on using virgin plastic in packaging is in the process of being implemented to "reduce packaging waste and stimulate Europe's transition towards a circular economy [66]. It is estimated that upon implementation, this initiative will generate €7 billion annually in revenue. A recent study by the

Institute of Australia [67], suggests that adopting a similar approach in Australia could yield $1.5 billion in revenue.

4.4. Financial Incentives for Procurement of PwRC

This policy refers to a monetary or economic benefit provided to individuals, businesses or organisations to encourage the use of PwRC. One of the major sources of financial incentives can be the hypothecated revenue that emerges from imposing a tax on waste disposal activities [68]. In research conducted in Australia [69], the results of a survey responded to by key stakeholders in the building and construction sector revealed that financial incentives are among the top ten factors that enable CE in that sector. Recently, in Western Australia (WA), the government issued a policy [70] that supports the local governments to source PwRC from accredited suppliers and use them in construction projects by providing a payment of $5 per tonne of PwRC. The previous research indicates that financial incentives are particularly effective in developing markets for PwRC [71, 72].

However, there is evidence from previous research that argues that financial incentives are not necessarily a major factor in encouraging key stakeholders to use more PwRC in construction projects. For instance, the findings from a recent research study [73] showed that this policy is barely a motive for using PwRC in construction projects. Instead, the participants in this research indicated that the environmental benefits, ensuring a competitive advantage and future proofing resulting from PwRC application play a more important role. Similarly, Calvo, Varela-Candamio [74] indicated that Spain's C&D waste recovery targets could be achieved in ten years using landfill levies, which was estimated to be faster compared to 12 years using financial incentives. In research conducted in 2020 [34], participants were asked to identify the key initiatives they believed the Australian federal government should prioritise to establish effective C&D waste management; the findings revealed that the GST subsidy for PwRC was perceived as the least important factor, with over 43% of participants expressing some level of disagreement with the idea of the government pursuing this strategy. This perception could be linked to the perceived risks associated with the application of these resources in construction projects.

4.5. Harmonised Recycled Product Specifications

In Australia, the use of PwRC in infrastructure construction projects is subject to compliance with requirements stipulated in recycled product specifications. These specifications are sometimes individually developed and administrated in states and territories. A recycled product specification is a document that outlines specific material types, qualities and dimensions required for PwRC. For instance, in New South Wales (NSW), there are several specifications [75] available for utilising particular PwRC in the construction of infrastructure and products. Figure 8 shows the public agencies involved in developing PwRC specifications [76] across eight states and territories in Australia.

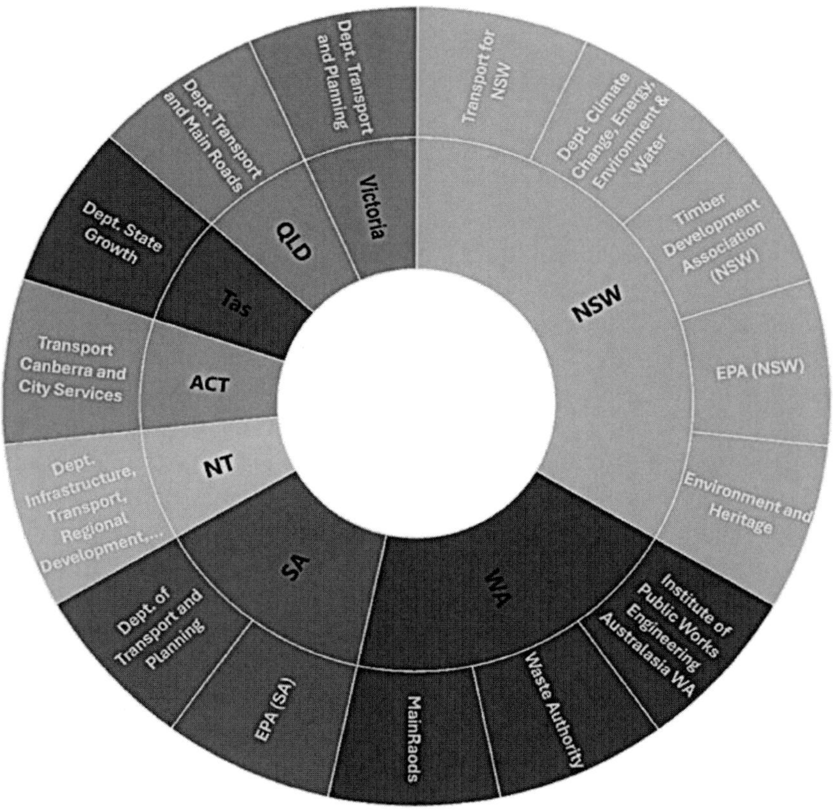

Source: Authors.

Figure 8. Public agencies that are responsible for developing PwRC specifications.

The fact that these documents are not primarily nationally developed or imposed is considered to be a major hurdle for those organisations that operate interstate where required. This approach can add complexity and cost to the operation and manufacturing of PwRC and limit the markets for these products as they need to be prepared according to various state-based specifications. In the presence of various recycled product specifications [76], following different testing criteria is necessary, which is both complex and costly. In research published in 2022 [34], perhaps the first step towards improvements is to harmonise the specifications documents. Any inconsistencies in these documents can significantly impact national efforts towards developing a market for PwRC in a negative way (p. 4326).

Hence, having harmony in these specifications can improve the PwRC application usability and competitiveness in the Australian market. Such harmonised recycled product specifications policies should allow a wide range of PwRC applications in civil works and infrastructure projects. In a report published by Equilibrium [61], the Australian industry's reaction to the potential development and implementation of new standards and specifications was extremely positive. Furthermore, national specifications were viewed as highly desirable outcomes, signalling a strong endorsement from the industry. However, due to the number and diversity of organisations involved in the process, it is not a straightforward task. The federal government has a leadership role in coordinating efforts to make sure consistent criteria for PwRC are applied across Australia.

4.6. Environmental Sustainability Rating

This policy is based on a systematic framework to assess, measure and rate the environmental sustainability of buildings, products, services and organisations in the building and construction sector. In Australia, there are a number of organisations that provide such ratings for construction projects. The most important national rating schemes include Green Star provided by the Green Building Council of Australia (GBCA) and NABERS by the NSW Government for building projects, IS rating [77] by the Infrastructure Sustainability Council (ISC) for infrastructure projects and EnviroDevelopment [78] by the Urban Development Institute of Australia for development projects.

Furthermore, several international rating schemes are also used in Australia. These include Living Building Challenge® by the International

Living Future Institute [79], BREEM by Building Research Establishment (UK) and LEED by the US Green Building Council. Table 5 summarises some global sustainability rating schemes in the building and construction sector.

Table 5. Some sustainability rating schemes from around the world

Name	Organisation	Description
• LEED	US Green Building Council (USGBC)	It provides a framework for green building design, construction, operation, and maintenance. It awards points for the use of PwRC in construction.
• BREEAM	UK Building Research Establishment	BREEAM assesses the environmental performance of buildings. It encourages the use of PwRC and awards credits for their incorporation.
• Green Star	Australia GBCA	It includes credits for the use of PwRC in construction
• Estidama	UAE Department of Urban Planning and Municipalities	Estidama promotes sustainable development through its Pearl Rating System. It encourages the use of PwRC and considers them in its assessment criteria.
• CASBEE (Comprehensive Assessment System for Built Environment Efficiency)	Japan Sustainable Building Consortium	CASBEE evaluates and rates the environmental performance of buildings and promotes the use of PwRC
• DGNB (Deutsche Gesellschaft für Nachhaltiges Bauen)	German Sustainable Building Council	It offers a certification system for sustainable buildings. It includes criteria for the use of PwRC.
• Green Building Index (GBI)	Malaysian Institute of Architects (PAM) and the Association of Consulting Engineers Malaysia (ACEM)	GBI encourages the use of PwRC as part of its assessment criteria
• IS (Infrastructure Sustainability)	Australia Infrastructure Sustainability Council	It evaluates the sustainability performance of infrastructure projects and considers factors such as resource efficiency, environmental impact, and innovation, including the use of recycled materials.
• UDIA	Urban Development Institute of Australia	It advocates for sustainable development practices, which may include the use of PwRC in urban development projects.
• Living Building Challenge®	International Living Future Institute	Petal Certification System awards the use of PwRC in building projects.

Source: Authors.

These rating schemes are based on credit systems that technically promote the efficient use of construction materials. For instance, under the Green Star scheme, construction projects can obtain credit points for waste-related practices, which include: the use of PwRC, and waste storage that promotes reusing and recycling of C&D waste from the project.

Previous research called for a reasonable set-up for specifying, testing and auditing the use of PwRC in construction projects through these rating schemes in Australia [80]. Australian governments have a key role in the wide adoption of this policy within the building and construction sector through enforcement or encouragement. For instance, from May 2024, achieving an energy efficiency rating of 7 has become mandatory for new housing in Australia [81], as stipulated in National Construction Code (NCC) volume two [82]. In 2023, Sustainability Victoria launched a program called 7 Star Homes to build capacity within the residential building market in preparation for an increase to this new minimum energy efficiency standards [83]. This program offers a range of capability-building opportunities such as rebates, as-built verification assessment, and training in designing and implementing effective marketing strategies.

4.7. Carbon Tax Scheme

A carbon tax scheme is an economic mechanism implemented by governments or regulatory bodies to incentivise and promote the use of PwRC in construction projects while simultaneously imposing a tax or fee based on the carbon emissions associated with using virgin materials. There is no mandatory carbon tax scheme currently being applied in Australia. The issue of carbon emission with regard to the production of PwRC is complex. On the one hand, carbon emissions during recycling are inevitable; however, the measures for substantial emissions are possible and crucial. On the other hand, the options of landfill disposal and extraction of virgin materials have far more adverse environmental consequences [84].

There are two widely adopted approaches to addressing the issues with emissions: Command Control (for example, through regulations, direct and indirect taxes) and a carbon tax scheme. The first approach comes with some limitations, as the unit cost for removing additional quantities of pollution is unreasonably expensive in some scenarios. Another issue with the Command Control approach is that it is stricter than the carbon tax scheme approach which is a more incentive-based system. In command control, the emission goal set for each polluter is fixed; hence, shifting the burden of pollution reduction to an organisation that can achieve it more cheaply is not possible. Thus, this approach is likely to be more costly in general [85] and, in most cases, the additional costs would be transferred to end users [86]; therefore, it is a less favoured method in circular-based systems.

On the contrary, implementing a carbon tax scheme can reduce emissions from waste disposal and recovery facilities. This is a market-driven approach to managing pollution by providing economic incentives to reduce the emission of pollutants [87]. In the general context, the idea is to mitigate the adverse effects of climate change and improve environmental protection. In the CE context, this scheme can convince waste producers to consider the top layers of the waste hierarchy (re-use, recycle and recover).

A version of the carbon tax scheme is an emission trading system (ETS). Through this scheme, the relevant authority allocates a limited number of permits to dispose of a certain amount of a specific pollutant during the period stipulated [88]. Polluters (waste producers) need to own permits in an amount equal to their emissions. Polluters who wish to add to their emissions should purchase permits from others willing to sell them. An ETS allows for emission goals to be met most cost-effectively by letting the market determine the lowest-cost pollution abatement opportunities. There are four main types of ETS: 'cap-and-trade system,' 'baseline-and-credit schemes,' 'project-based schemes' and 'hybrid schemes.' The EU implemented a cap-and-trade system in 2005 [89] under the Kyoto Protocol and aims to reduce GGE in an economically effective manner.

The ETS in Australia has been a source of disagreement between the major political parties because of its social and economic effects. Between 2003 and 2011, policies related to the ETS were passed and overturned several times. The Australian Parliament has provided a list of the pros and cons of the ETS concerning varying factors [90]. The first ETS in Australia was established in NSW in 2003, based on a baseline-and-credit scheme [91]; it only lasted for a decade and was terminated in 2012. A report [92] that reviewed this scheme's performance indicated a high level of commitment from different stakeholders during its lifetime.

Nevertheless, around five million certificates remain available for voluntary surrender on this scheme registry. Later in 2012, the Australian Government initiated a carbon pricing scheme, or carbon tax, through the Clean Energy Act 2011. The purpose of this Act was to make polluters pay a certain amount ($23) as tax per tonne of carbon that they released into the atmosphere. However, this Act was repealed in 2014 and replaced with the

Direct-Action Plan, which provides funding to companies to incentivise emission reduction activities. The government has spent $1.7 billion on 143 million tonnes of emissions, at an average cost of $12 per tonne. This fund is granted on a ‘reverse auction’ basis, awarding contracts to those who bid on emissions abatement projects at the lowest cost.

The Australian Government has committed to a target of GGE abatement by 26- 28% (from 2005 levels) before 2030. Some state waste strategy documents prioritise emissions reduction through increased waste recovery activities. South Australia (SA) is the first Australian jurisdiction to enact specific climate change legislation that sets an ambitious long-term emissions reduction target. Through the Climate Change and Greenhouse Emissions Reduction Act 2007, SA establishes a target to reduce its GGE by at least 60% (from 1990 levels), by 2050. In the Australian Capital Territory (ACT), the Climate Change and Greenhouse Gas Reduction Act 2010 has provided a target of 40% emissions by 2020, while the waste sector only accounts for 3% of total emissions. In Queensland (Qld), an environmental strategy document necessitates implementing an ETS to reach a 60% target of reduction in national GGE by 2050.

There is some uncertainty about how a carbon tax, ETS or reverse auction may impact the waste recovery sector in Australia. In a previous report [93], consultation with waste recyclers revealed that introducing the carbon tax may result in more emphasis on the recovery of C&D waste, as landfill operators should report on and pay a price for their activities-produced emissions. Several studies have also compared the effectiveness of the two GHG management mechanisms. One modelling study in 2014 showed that an ETS can reduce GGE from waste by 75.9% (from 2015 levels) by 2030 [94]; the study, however, stated that an ETS is likely to reduce Australia’s gross domestic product (GDP) by just over 1.1% in 2030 compared to a base case.

In 2016, the findings of a study [95] demonstrated that ‘direct auction’ was not as effective as a carbon tax in forcing companies to act urgently and manage emissions. The interviewees in this piece of research stated that the carbon tax motivated companies to act, as it raised their utility costs, causing a financial burden for some companies and ruining their reputation as high-emitting companies, in addition to these companies being liable under the tax. The study also indicated that the focus on carbon emissions in these companies shifted when the carbon tax was repealed. Another piece of research [96] compared three models of GGE reductions (ETS, auction plan and carbon tax) and found that an ETS is the most viable option for reductions in GGE and economic growth. This research predicts that the government will encounter much higher prices in the next rounds of auctions than in previous auctions. Hence, the current budget ($2.55 billion) may not be sufficient to purchase the required abatement by 2020, making achieving the 2030 target difficult. Another study in Victoria [84] warned that an ETS might also cause the industry to rethink recycling emissions levels and this would negatively impact further recycling activities. In general, and in the interest of CE and the resource recovery industry, legislation should be modified to account for the fuel tax cuts.

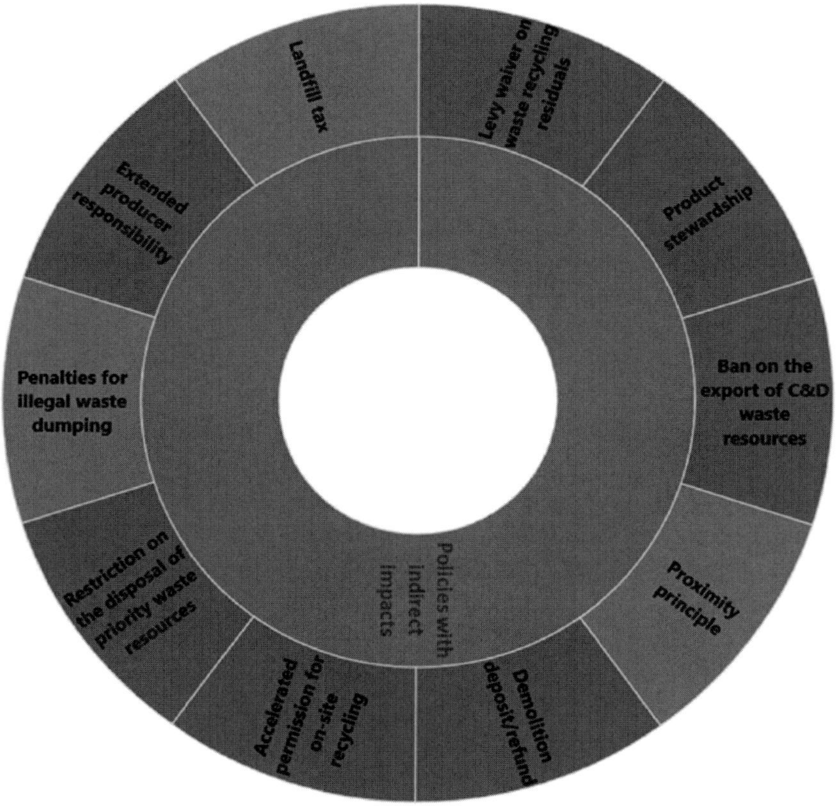

Source: Authors.

Figure 9. Policies with direct impacts on PwRC application. Green denotes current Australian policies; grey denotes non-current policies.

5. Policies with Indirect Impacts

Figure 9 presents the visual representation of the CE policies identified in this review, which are considered to have an indirect impact on the adoption of PwRC in construction projects. The following sections will provide a thorough exploration of these policies.

5.1. Landfill Tax

The building and construction sector's current preferred method for dealing with C&D waste is landfilling. This is because of its convenience and cheap cost (via low landfill levies) compared with alternative waste management methods, as well as its accessible hours of operation [97, 98]. Udawatta, Zuo [99] contend that poor financial returns and lack of incentives inhibit subcontractors' interest in waste minimisation. Newaz, Davis [100] also point this out, arguing that construction projects' variety makes consideration of waste management difficult during the design phase. Analysis of numerous fit-out case studies in NSW indicates that 78% of the waste produced on-site goes to landfill [97]. Recycling facility fees applied to waste materials processing compared to those at landfill sites are greater [101]. Ratnasabapathy, Perera [102] demonstrate in a recent study that a minor reduction in Australian C&D waste disposal highlights the necessity of devising ways to limit landfilling. The researchers predict that, if the trajectory continues, this will result in a 78% diversion rate from landfill by 2025.

Most of the existing studies on this topic emphasise the need for landfill levies. For example, studies suggest that the recovery of waste can be driven by landfill levy fees that are high. Shooshtarian, Maqsood [103] state that, of participants in one survey, 90% recommended landfill levies as generally effective. Landfill levy rates in Australia are decided on by state and territory governments [41]. This means there is variation between states, depending on how the rates are formulated. A study of states' recent waste strategies found that these documents recommend the revision of current levy practices to dissuade the industry from waste disposal as their second-highest suggestion [104]. The jurisdictions of ACT, SA, Qld, Tasmania (Tas) and WA all suggested this way forward.

Research has been conducted into landfill levies' standards in Australia as well as their suitability [95, 105]. Stakeholders are concerned about the effective design of such a scheme to better mirror practical considerations plus expand its influence [106]. Shooshtarian, Maqsood [68] contend that while landfill levies may be the preferred market approach in certain scenarios, they can also inadvertently discourage best waste management practices. For example, these levies may lead to increased illegal dumping, shifting of waste to other regions and a reduction in overall recycling efforts. Rameezdeen, Chileshe [107] research into SA's construction sector, with regards to RL, shows that illegal dumping can be an unintended consequence of a higher landfill levy rate. In WA, a government document as cited in Zhao, Webber [106] echoes this finding. Figure 10 displays where in a material's lifecycle these policies can be implemented.

A Qld study showed how the state government decisions regarding the establishment and rescinding of landfill levies can impact the C&D waste recycling rate across the state [108]. A study by Wu, Zuo [109] showed that there are some levels of waste transfer between a number of Australian states e.g., NSW and ACT, NSW and Qld and SA and Victoria. The researchers indicated that the recycling/landfilling site availability and inconsistent levy rates are the main reasons for these inter-state waste transports. This issue can be minimised by implementing policies such as the proximity principle [105], in which waste transfer is restricted over long distances. Newaz, Davis [100] report specialists' opinions that NSW's levy on waste is ineffective in terms of minimising waste landfilling, due to there being few practical results from levies and the fact that the state permits waste transfer to Qld, which has comparatively lower landfill levies. In Shooshtarian, Maqsood [68], participants report that the current landfill levy schemes used in Australia could be more efficient and require changes to fulfill waste diversion goals.

Source: Authors.

Figure 10. The application of studied policies to the construction material lifecycle.

Generally, landfill levies are determined based on the location of landfills, levy exceptions for particular materials, how waste is composed and levy zones. For this reason, they rise regularly [72, 98]. Jayasinghe, Rameezdeen [105] argue that levies should be decided depending on the waste classification, demographics, financial impact analysis and levy rebates for waste recovery. Landfill levy revenues are in part spent on bolstering enforcement and compliance, devising good policies, and funding initiatives and activities that foster waste minimisation. However, Australia lacks a national approach to the spending of levies on the aforementioned purposes, with each state government doing so as guided by its specific concerns and preferences [68].

Several studies propose strategies targeting waste landfilling and landfill levies. Shooshtarian, Maqsood [68] contend that strategies to enhance landfill levy effectiveness include consideration of fees on transport and possible price repercussions for the building and construction sector when deciding the levy fees; standardisation of levies for landfill; supplementing levies with technology-driven compliance and enforcement systems; improving local infrastructures; directing revenue from landfill levies into resource recovery via initiatives such as offering monetary incentives or low-interest loans; and funding R&D activities. A good strategy for better resource circularity in landfills is to increase the use of energy recovered (mainly methane gas) from combustible C&D waste materials that would otherwise end up in landfills [110].

Tam, Kotrayothar [101] suggest that landfill operators should use differential applications of a levy to encourage C&D waste source separation. Additionally, competent waste data and reporting systems that enumerate the precise quantity of landfilled waste would enhance planning and regulation for greater waste minimisation via landfilling or otherwise [111]. Effective waste landfilling is complemented by locating engineered landfills close to transfer stations and reprocessing centres for optimal transportation [105]. A combination of properly designed landfill levies, the proximity principle and locating landfill sites at further distances, could encourage recycling instead of landfilling.

5.2. Extended Producer Responsibility

Extended producer responsibility (EPR), also called the "polluter pays" or "take-back" principle, is a policy approach designed to encourage manufacturers, suppliers and importers of construction materials to take active responsibility for waste minimisation. EPR makes the material's manufacturer or its supplier accountable - physically and/or monetarily - for the entire lifecycle of their materials, including the waste generated from the application of a product [112].

When effectively implemented, EPR can motivate producers to proactively address the environmental impacts of their products. This approach encourages waste prevention at the source by promoting sustainable product design, technological advancements and the adoption of eco-friendly designs. Additionally, EPR fosters efficient waste management strategies within large-scale production planning, creating a ripple effect that contributes to minimising waste generation across the entire production lifecycle.

5.2.1. Criteria Used to Develop EPR Policies

Previous studies have sought to develop models that address the intricate factors influencing EPR to improve its practical application and outcomes [113]. For instance, Dubois, Graaf [1] offered five measures of the capability of EPR for C&D waste management development and assessment, as shown in Figure 11a. The authors argued that the EPR application in the study context is primarily driven by two key factors: environmental scope and political priorities. Furthermore, Acree Guggemos and Horvath [2] put forward a policy framework to optimise the achievement of EPR goals for C&D waste management. This framework which is based on Thorpe and Kruszewska's (1999) [114] model, comprises three forms of EPR policy instrument: regulatory, economic and information-based (Figure 11- right).

Other studies have also suggested alternatives that involve a few overlaps and divergences compared to the models presented in Figure 11 [115-119]. Furthermore, some studies present valuable insights into optimising EPR policies by examining factors influencing their outcomes. For instance, Gupt and Sahay (2015) [120] conducted a comparative analysis of 26 case studies to find the elements that are linked with successful EPR applications and the key aspects of EPR expansion and execution. They found that 'recycling agencies' and 'source segregation' as well as 'financial responsibility of the producers' strongly support positive EPR outcomes. A study in 2016 [121] on the effectiveness of various environmental policies revealed that EPR policies are considered to be highly effective but that there is uncertainty about their acceptability. The following section provides insights into the challenges of EPR policy development and application.

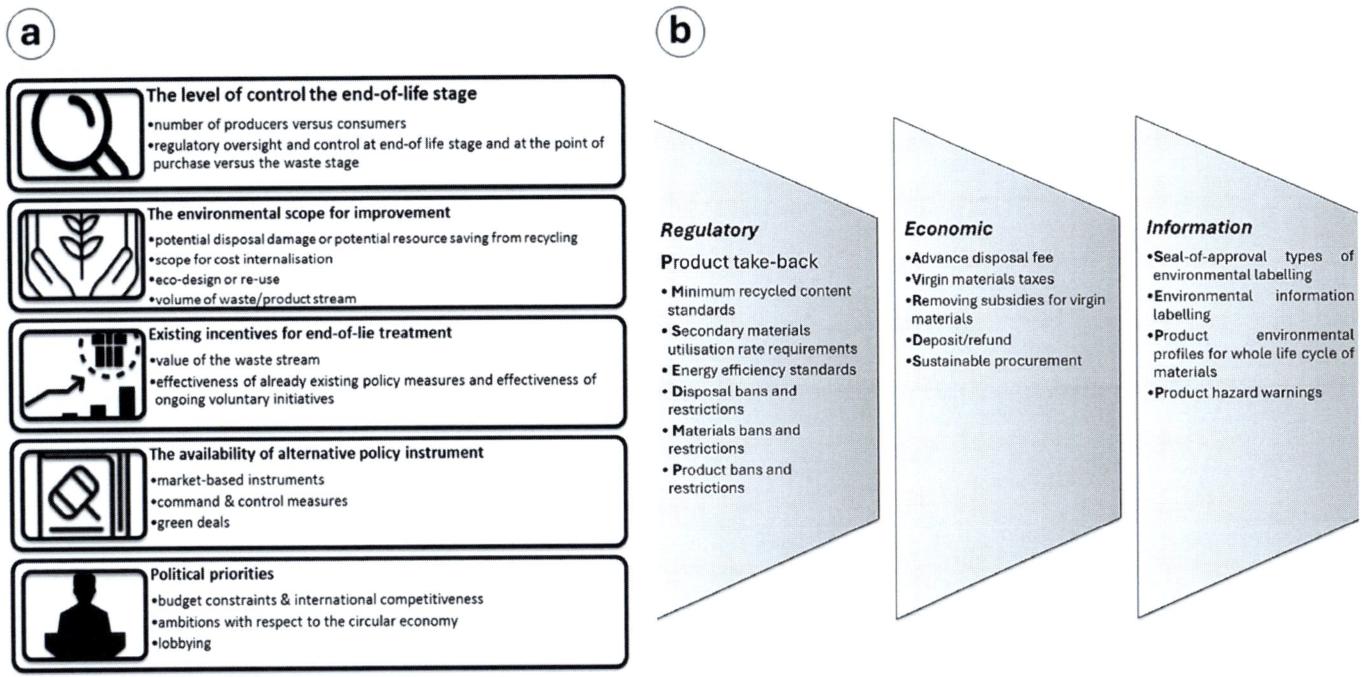

Source: Adapted from Dubois, Graaf [1] and Acree Guggemos and Horvath [2].

Figure 11. Five criteria suggested for the development of EPR policy (a). Three policy instruments that facilitate EPR implementation (b).

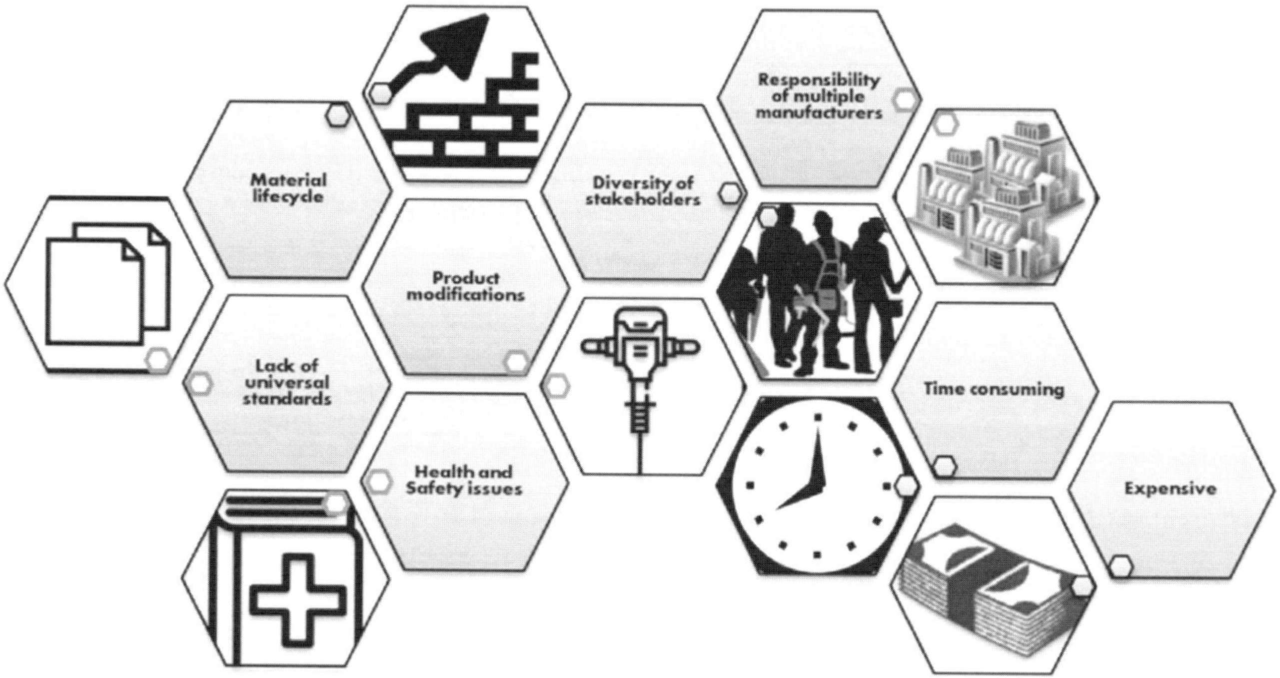

Source: Shooshtarian, Maqsood [124].

Figure 12. Key obstacles limiting the successful implementation of EPR in the built environment.

5.2.2. The Main Challenges to Adopting EPR in the Building and Construction Sector

The building and construction sector's widespread embrace of EPR policies is held back by a number of obstacles. These challenges are outlined in Figure 12 [2, 122, 123]. Obstacles to establishing EPR policies include cost and time implications; complexity of policy establishment and enforcement; responsibility of manufacturers; the range of stakeholders involved in the process; the lifecycle of materials in construction; and OH&S to address these issues, suggestions include the establishment of an effective supply chain system; encouraging Design for Deconstruction (DfD); product documentation; waste responsibility assignment; and enhanced health and safety risk management.

5.2.3. The Main Challenges to Adopting EPR in the Australian Building and Construction Sector

Implementing EPR in Australia's building and construction sector presents significant challenges due to the diverse range of stakeholders involved in the lifecycle of construction materials. These stakeholders include manufacturers, suppliers, contractors and waste management entities, with each playing a role in the production, distribution, use and disposal of construction products. The absence of a unified and comprehensive EPR strategy in this sector is largely attributed to the complexity of coordinating these varied participants. This complexity is further compounded by the need to align differing interests, regulatory frameworks and operational practices across the sector. Consequently, developing and implementing a harmonised EPR approach requires extensive collaboration and integration among all parties to effectively minimise waste and promote sustainability within the sector [124].

There is no mandatory EPR policy legislated in Australia at present for construction material manufacturing. NSW and WA, however, do have voluntary EPR policies [125]. Two studies in Australia [33, 126] point to the lack of a mandatory EPR policy as an obstacle to the reuse of waste materials and the creation of C&D waste trading markets in Australia. Shooshtarian, Maqsood [124] studied the implementation of EPR in Australia and found that there is significant agreement between various stakeholders about the need for EPR.

5.3. Penalties for Illegal Waste Dumping

In the Australian C&D waste management environment, illegal dumping and long-term stockpiling pose significant challenges. While the National Waste Report [3] suggests that only 1% of C&D waste is from illegal dumping, anecdotal evidence suggests a larger percentage, especially in the Northern Territory (NT) and WA. The existing literature finds a range of motives for illegal dumping and stockpiling. For example, a study by Shooshtarian [10] suggests that the presence of varying jurisdictional regulatory frameworks across Australia contributes to illegal dumping and long-term stockpiling. In two other studies, Rameezdeen, Chileshe [107] and Shooshtarian, Maqsood [68] find that higher landfill levy rates can cause illegal dumping. Shooshtarian, Maqsood [127] acknowledge the limited return of levy revenue to the C&D waste management sector as a barrier to effectively preventing illegal waste practices.

Australian states and territories address illegal dumping with substantial fines enforced by specialised task forces, coupled with ambitious targets aimed at reducing these practices [128]. However, penalties for illegal waste dumping vary across Australia's jurisdictions, with the highest being imposed in NSW, where offenders face up to $5M and/or seven year imprisonment (for wilful offences). In NT, Tas and ACT, the penalty is $5M (+/five years imprisonment), $1.59M (+/five year imprisonment) and $1M (+/seven year imprisonment), respectively. This is followed by SA (up to $30K), WA (up to $125K), Qld (up to $217K) and Vic ($775K).

The state governments use waste strategy documents to guide efforts to reduce illegal dumping [104], The overarching structure to guide these efforts includes education (awareness-raising, stakeholder engagement); enforcement (evaluation and monitoring, regulations); and encouragement (infrastructure development, capacity building and networking).

The specific strategies to reduce illegal dumping and stockpiling include enhanced supervision by demolition companies to deter improper disposal [129]; implementing standardised regulations with uniform levy fees [38, 130] and allocating funds for educational programs [38]. Furthermore, adopting advanced technologies such as image processing, remote sensing and GIS [131, 132] can provide effective oversight, enabling authorities to monitor and control illegal waste activities more efficiently. Lastly, for the prevention of long-term stockpiling, an upfront levy obligation can be implemented Jayasinghe, Rameezdeen [105]. This approach applies a fee to waste upon arrival at disposal depots, incentivising quicker processing and reducing the likelihood of prolonged stockpiling. This method is currently practised in NSW.

5.4. Restriction on the Disposal of Priority Waste Resources

Restricting the disposal of priority waste resources can incentivise the waste recovery sector by guaranteeing a consistent and sustainable supply of feedstock for facilities processing C&D waste. This approach not only supports the viability of recycling operations but also promotes the efficient use of waste materials that would otherwise end up in landfills.

In the Australian waste regulatory environment, priority waste resources are those with high disposal impacts (such as toxicity or greenhouse gas emissions), social impacts (such as community concern or amenity) or whose recovery would present resource savings or business opportunities [133]. However, not many C&D waste materials are classified as priority waste in Australian state and territory waste regulatory frameworks. For instance, in SA, of 21 waste materials banned from disposal at landfilling, only Polyvinyl Chloride (PVC) and aggregated metals fall under the C&D waste stream. In Vic, WA and Qld e-waste is banned from landfilling. In NSW, there is currently only one program aimed at banning household organics from landfills, which is scheduled to be implemented in 2030.

The disposal ban approach falls under three broad definitions of waste by source, by type and by property. In Europe, there have been a number of bans on, for example, unsorted waste, untreated waste, treated and/or untreated organic wastes (with specific properties) and combustible waste. The bans are defined by a mixture of all three waste categories [134].

5.5. Accelerated Permission for On-Site Recycling

This policy is a regulatory or administrative process designed to fast-track the approval of temporary waste recovery facilities that are conducted on-site. Accelerated permission for on-site recycling can facilitate the uptake of PwRC in the building and construction sector. On-site recycling typically offers greater environmental and economic benefits compared to off-site recycling. This is because transporting waste resources between the construction site and the recycling facility incurs costs and emissions.

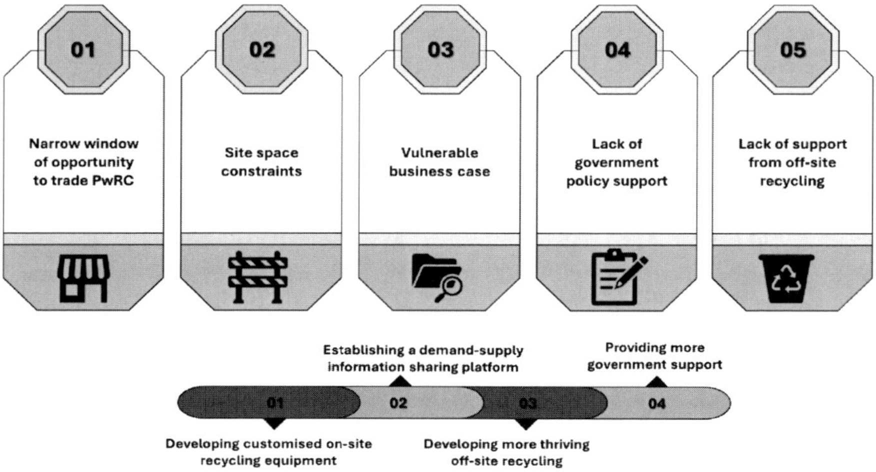

Source: Adapted from Bao et al., (2020) [137].

Figure 13. Major barriers and facilitators of on-site recycling.

In a research project concluded in 2023 [135], it was found that the absence of this policy poses a significant challenge to reusing PwRC in construction projects. For instance, in a government-led residential development project in WA, the initial plan was to recover demolition waste on-site and utilise the resulting PwRC. However, securing approval for on-site recycling took almost 12 months, which significantly prolonged the project's timeline [136]. Currently, no state in Australia offers fast-tracked permission for on-site recycling of C&D waste resources. One study in Hong Kong [137] identified the key barriers and facilitators of on-site recycling for construction and demolition projects (Figure 13). One key barrier was found to be 'lack of government policy support' which was further suggested to be addressed by 'providing more government support.'

This policy has been successfully implemented overseas for several years. For example, in certain Canadian cities, developers are required to pay a refundable fee to obtain demolition permits. This fee is reimbursed once they demonstrate that at least 75% of their demolition waste has been recycled [138]. Various models have been proposed to illustrate the application of this policy in the building and construction sector [74, 139]. For instance, Liu, Teng [139] introduced a six-step model that provides a structured approach to implementing the policy:

1. Firstly, the project owner (developer) estimates the C&D waste of a project and then submits the waste assessment report to the waste regulatory body which will inform the owners to submit a deposit. The amount of the payment depends on the type of project involved and the waste assessment report.
2. The owner submits a pre-deposit, and the body issues a construction permit for the construction or demolition of the project.
3. Based on the waste assessment report, the owner will make a construction waste management plan in person, or by entrusting a contractor, specialised in resource recovery, to be then implemented by a contractor.
4. Contractors treat the C&D waste according to the waste management plan, such as transporting waste materials to recycling sites or conducting on-site recycling.
5. After the completion of the project, the Waste Assessment Centre will evaluate the waste treatment of the project and decide whether or not to issue the waste disposal certificate to the owner and contractor depending on the assessment results. The Centre is an organisation belonging to the waste regulatory body.
6. The body then returns the pre-deposit to the developer. In the meantime, it will make cost compensation or reduce the tax to the contractor. If the compensation is taken, the amount of compensation will be multiplied by the appropriate proportion based on the construction waste estimate report and the disposal cost.

5.7. Proximity Principle

The underlying concept of this policy is to compel companies that create waste to deliver it to a location for processing that is within a particular distance from where it originated. The application of the proximity principle (PP) can help stop waste transport from one location to another to avoid and minimise levy liabilities. The need for implementation of this policy was highlighted in submissions to the Environment and Communications References Committee (2018) [130]. In NSW, the PP was introduced in 2014 to limit the transportation of waste over long distances, aiming to reduce environmental impacts and encourage local waste management. This regulation prohibited transporting waste more than 150 kilometres from its origin unless it was to one of the two nearest lawful disposal facilities. The Waste Contractors and

Recyclers Association of NSW argues that the adoption of a nation-wide PP would lead to better outcomes. In turn, the Law Council of Australia (LCA) argues that any nation wide PP must consider the Constitution's Section 92 stipulation that commerce and trade between Australian states must be completely free.

Meanwhile, the Waste Management Association of Australia (WMRR) submits that long-distance waste transportation must cease, stating that …*we do not agree with long-distance transportation; we actually agree there has to be a proximity principle in place to stop the excessive and unnecessary movement of waste across distances, particularly if there is the infrastructure in place. You can't actually invest and develop infrastructure if you haven't got certainty about what's coming through the front gate. In Europe, you do have a proximity principle, so we need to solve how we do that (p. 59).*

Another contentious aspect of the PP is the need for authorities to carefully consider its broader environmental impacts. Critics argue that while the PP aims to reduce waste transport distances, the environmental benefits may be undermined if local facilities lack the capacity or technology for effective processing. In some cases, limiting transport could lead to inefficient waste treatment or increased reliance on landfills, which may offset the intended environmental gains of the policy. Furthermore, it is argued that market development for PwRC may rely on the exchange of waste materials (both recovered and unrecovered) across various regions as it supports and maintains sectors and companies within the economy.

To address the impact of this policy on the Australian context, the government and the waste recovery industry have pledged that they will create an effective, long-term national market for the trade of Australian waste products [140]. To address the impact of this policy on the Australian context, the government and the waste recovery industry have pledged that they will create a robust, effective, self-sustaining, long-term national market for the management and trade of Australian waste products domestically. PP policy objectives align with these objectives by encouraging shorter, more efficient waste transport and minimising environmental impact, supporting Australia's transition towards CE.

5.8. Ban on the Export of C&D Waste Resources

The policy involves bans implemented by a government or an authority to prohibit or restrict the export of C&D waste materials to other countries. In

Australia, the regulatory responsibility for exporting waste materials was shifted from the state level to the national level in 2020. This year, Australian governments agreed that the export of some waste materials (tyre, plastic, glass, paper and cardboard) would be regulated by the federal government [141]. Following this agreement, the Recycling and Waste Reduction Act 2020123 was developed to provide statutory power to waste export ban. This Act, however, does not include a ban on the export OF C&D waste resources.

Nevertheless, glass, plastic and cardboard arising from C&D activities are subject to this regulation. For instance, this includes materials like plastic or cardboard used in packaging construction materials. An export ban can serve as a significant incentive for recycling, prompting further investment in waste recovery facilities and technologies. This, in turn, enhances the quality and cost-effectiveness of PwRC, leading to its optimal adoption within the sector. It also instils confidence in the waste recovery industry, ensuring that sustainable feedstock is available for waste recovery facilities. The findings from a survey [142] revealed that only 23.5% of stakeholders indicated that the ban on waste imports by foreign countries (e.g., China, Malaysia and Thailand) had a negative impact on Australian C&D waste management.

This policy should be pursued in conjunction with capacity building in the waste recovery industry and the stimulation of markets for PwRC. Otherwise, it could lead to the accumulation of large stockpiles of recyclables or PwRC, which can pose various risks, including the risk of fire [143].

5.9. Product Stewardship

Product stewardship acknowledges those involved in designing, manufacturing and selling products have a responsibility to ensure those products or materials are managed in a way that reduces their environmental and human health impacts, throughout the product's life cycle and across the supply chain. It aims to drive environmentally beneficial outcomes through good design and clean manufacturing, including the use of components and materials that are easier to recover, reuse and recycle. Developing product stewardship schemes targeting construction materials supports the production of PwRC and contributes to the optimal uptake of these resources in construction projects.

Table 6. Current product stewardship schemes in Australia

Product	Arrangement	Description
Oil	Mandatory	The Government provides industry incentives to increase recycling of used oil.
TVs and computers	Co-regulatory	The National Television and Computer Recycling Scheme (NTCRS) was established in 2011. It gives Australian households and small businesses free access to industry-funded collection and recycling services
Plastics and packaging	Co-regulatory	The Australian Packaging Covenant (APC) [145] requires companies to minimise their packaging waste
Mobile phones	Voluntary	MobileMuster collects and recycles mobile phones, batteries, modems and accessories. This includes running public awareness campaigns to increase uptake in recycling of unused and damaged mobile phones.
Tyres	Voluntary	Tyre Stewardship Australia manage the national Tyre Product Stewardship Scheme. The scheme promotes the development of viable markets for tyre-derived products.
Large plastic bags	Voluntary	Big bag recovery collects and recycles bags used for products over 15 kg or 15 L that would otherwise be sent to landfills.
Batteries	Voluntary	The Battery Stewardship Scheme B-cycle is a national battery collection network that recovers and recycles used hand-held batteries. It also raises public awareness of battery safety and disposal options.
Aluminium cladding	Voluntary	The Fairview aluminium cladding scheme – Ecoloop takes used aluminium cladding and recycles it into new materials for local manufacturers.
Plastic paint pails	Voluntary	Dulux Project Earth recycling and remanufacturing of used Dulux brand plastic paint pails from commercial and industrial work sites.

Product stewardship schemes may be voluntary (industry-led), mandatory (under law) or shared with the industry (co-regulatory arrangements between government and industry) [144]. Currently, in Australia, there are three active product stewardship schemes (Table 6).

Companies administrator in a voluntary product stewardship arrangement can apply to be accredited by the Australian government [146]. This accreditation program enables companies to demonstrate to businesses and consumers that the arrangement's environmental and human health outcomes have been verified and will contribute to Australia's recycling and waste reduction objectives.

The Product Stewardship Centre of Excellence [147] was established in Australia to help businesses, industries and product stewardship organisations adopt a product stewardship approach to deliver their Environmental, Social, Governance (ESG) and CE objectives. This Centre is an independent, non-profit charity that provides independent advice on product stewardship priorities to the Australian Government. The 2023-24 priority product list [148] for product stewardship includes clothing textiles, tyres, plastic in health care products, mattresses and child car seats. The Centre's product stewardship assessment framework [149] comprises 33 key performance indicators (KPIs), which encompass the environmental, social and economic benefits of applying product stewardship. Several of these KPIs, including the use of PwRC, recovery rate, sustainable procurement and local recycling, directly contribute to maximising the uptake of PwRC.

It is noteworthy to know that Product Stewardship and EPR policies differ in their scope and responsibility level. Product Stewardship and EPR differ in their scope and responsibility. Product Stewardship involves shared responsibility among all stakeholders, including producers, consumers and waste managers, and it can be either voluntary or regulated. Its focus is broader, covering sustainable product design and lifecycle management. In contrast, EPR places the primary responsibility on producers, typically through mandatory legislation, to manage the end-of-life disposal and recycling of their products. While Product Stewardship fosters collaboration across the value chain, EPR aims to reduce government waste management burdens and incentivise producers to consider their products' end-of-life stage.

5.10. Levy Waiver on Waste Recycling Residuals

Recycling residual waste is a complex mix of materials that can be difficult to recover. This is because the mix requires expensive and complex infrastructure to recover. It is estimated that the residual waste from C&D waste recycling is around 20% [150]. The application of the levy waiver policy should ideally lead to greater C&D waste recovery and PwRC cost reduction and market development. Imposing a levy on the disposal of recycling residuals can significantly reduce PwRC's competitiveness in the market. By adding this cost, the economic advantage of recycling activities diminishes, making PwRC less appealing to stakeholders who might otherwise choose them over virgin alternatives.

A study in Victoria showed that for each $15/t levy rate rise, the steel recycling industry in this state sustained an extra $ 738k per annum cost [151]. At present, in all jurisdictions, waste recycling facilities' residuals are defined as waste, and therefore, the owners of such businesses incur a landfill fee. Recycling facilities face substantial expenses in disposing of residuals, naturally prompting them to explore lower-cost alternatives. This financial burden often justifies transporting waste to interstate locations with more favourable disposal fees. Consequently, the relevant industries suggest that these residuals be excluded from the waste definition by regulations; some examples include:

> A levy on the disposal of recycling residuals reduces the competitiveness of materials sold into the international market [National Waste and Recycling Industry Council]
> When recyclers are liable to pay the levy for the disposal of contaminants that have entered the recycling stream, they see it as a disincentive towards being involved in the recycling industry and instead, it encourages shipping unprocessed waste overseas [The Australian Council of Recycling]
> The disposal of residuals generally represents a significant cost for recycling facilities, which can obviously create commercial incentives to seek lower disposal cost options; it also justifies the transport of waste to interstate locations with a lower disposal rate [Re Group]
> Landfill levies penalise the recycling industry for the disposal of residual rubbish that enters the recycling stream [Visy, Owens-Illinois and SKM Recycling].

References

[1] Dubois, M., Graaf, D. D. and Thieren, J., *Exploration of the role of extended producer responsibility for the circular economy in the Netherlands*. 2016: London, UK.

[2] Acree Guggemos, A. and Horvath, A., Strategies of extended producer responsibility for buildings. *Journal of Infrastructure Systems,* 2003. 9(2): p. 65-74.

[3] Blue Environment, *National Waste Report*. 2023.

[4] Elmualim, A., Mostafa, S., Chileshe, N. and Rameezdeen, R., *Construction and the Circular Economy: Smart and Industrialised Prefabrication*, ed. R. Crocker, et al., 2018, Leeds: Emerald Publishing Limited.

[5] Guerra BC and Leite F, Circular economy in the construction industry: An overview of United States stakeholders' awareness, major challenges, and enablers. *Resources, Conservation and Recycling,* 2021. 170: p. 105617.

[6] Kirchherr, J., Reike, D. and Hekkert, M., Conceptualizing the circular economy: An analysis of 114 definitions. *Resources, Conservation & Recycling*, 2017. 127: p. 221-232.

[7] Anastasiades, K., Blom, J., Buyle, M. and Audenaert, A., Translating the circular economy to bridge construction: Lessons learnt from a critical literature review. *Renewable and Sustainable Energy Reviews,* 2020. 117: p. 109552.

[8] Parliament, E., *Circular economy: definition, importance and benefits*. 2022.

[9] Benachio, G. L. F., Freitas, M. D. C. D. and Tavares, S. F., Circular economy in the construction industry: A systematic literature review. *Journal of Cleaner Production*, 2020. 260: p. 121046.

[10] Shooshtarian, S., Maqsood, T., Wong, P. S. P., Khalfan, M. and Yang, R. J., Market development for a construction and demolition waste stream in Australia. *Journal of Construction Engineering, Management & Innovation*, 2020. 3: p. 220-231.

[11] Peck, P., Richter, J., L., and Delaney, K., Circular economy Sustainable materials management: A compendium by the international institute for industrial environmental economics (iiiee) at Lund University. 2020, *The International Institute for Industrial Environmental Economics (IIIEE): Lund, Sweden.*

[12] Shooshtarian, S., Maqsood, T., Wong, P. S., Yang, R. J. and Khalfan, M., Market development for construction and demolition waste stream in Australia. *Journal of Construction Engineering, Management & Innovation,* 2020. 3(3): p. 220-231.

[13] National Waste Policy, Less Waste. More Resources, W.a.t.E. *Department of Agriculture,* Editor. 2018: Canberra, Australia.

[14] Iyer-Raniga, U., Gajanayake, A. and Ho, O. T. K., The Transition to a Circular Built Environment in Australia: An Analysis of the Jurisdictional Policy Framework. *Environmental Policy and Law*, 2023. 53(4): p. 233-246.

[15] National Waste Policy Action Plan, 2018 National Waste Policy: From waste to resource, W.a.t.E. *Department of Agriculture,* Editor. 2019: Canberra, Australia

[16] Office of Energy and Climate Change, Circular design guidelines for the built environment, *NSW Government*. 2023, NSW Treasury: Sydney, Australia.

[17] Sterk, E., Incentives for construction clients in Germany to choose concrete with recycled aggregates, in *MAGKS Joint Discussion Paper Series in Economics*. 2023.

[18] Australia, C.o., *Sustainable Procurement Guide.* 2013.

[19] Alhola, K., Ryding, S. O., Salmenperä, H. and Busch, N. J., Exploiting the Potential of Public Procurement: Opportunities for Circular Economy: Public procurement promoting circular economy. *Journal of Industrial Ecology,* 2018. 23(5): p. 96-109.

[20] Brammer, S. and Walker, H., Sustainable procurement in the public sector: an international comparative study. *International Journal of Operations & Production Management*, 2011. 31(4): p. 452-476.

[21] Pick, F., *Waste as a resource: tools for construction and demolition waste management*. 2017.

[22] Craighill, A. and Powell, J., *A lifecycle assessment and evaluation of construction and demolition waste.* 1999.

[23] Zhang, C., et al., Upgrading construction and demolition waste management from downcycling to recycling in the Netherlands. *Journal of Cleaner Production*, 2020. 266: p. 121718.

[24] He, Q., Shi, S. and Hu, M. Exploring a Long-Term Mechanism of Construction and Demolition Waste Recycling: A Case of Chongqing. in *Proceedings of the 17th*

International Symposium on Advancement of Construction Management and Real Estate. 2014. Berlin, Heidelberg: Springer Berlin Heidelberg.

[25] Lingegård, S., Olsson, J. A., Kadefors, A. and Uppenberg, S., Sustainable Public Procurement in Large Infrastructure Projects - Policy Implementation for Carbon Emission Reductions. *Sustainability,* 2021. 13(20): p. 11182.

[26] Caldera, S., Mohamed, S. and Feng, Y., Evaluating the COVID-19 impacts on sustainable procurement: Experiences from the Australian built environment sector. *Sustainability,* 2022. 14(7): p. 4163.

[27] Ershadi, M., Jefferies, M., Davis, P. and Mojtahedi, M., Achieving Sustainable Procurement in Construction Projects: The Pivotal Role of a Project Management Office. *Construction Economics and Building,* 2021. 21(1).

[28] Caldera, S., Ryley, T. and Zatyko, N., Enablers and barriers for creating a marketplace for construction and demolition waste: A systematic literature review. *Sustainability,* 2020. 12(23): p. 9931.

[29] Sanchez, A. X., Lehtiranta, L., Hampson, K. D. and Kenley, R., Evaluation framework for green procurement in road construction. *Smart and Sustainable Built Environment,* 2014.

[30] Shooshtarian, S., Le, T., Feng, Y. and Bettini, L., Analysis of sustainable procurement in supplying recycled content: A case study in Western Australia in *World Building Conference 2022- Building our future: Informing practice to enhance the lives of current and future generations.* 2022, RMIT University: Melbourne, Australia.

[31] Shooshtarian, S., Maqsood, T., Caldera, S. and Ryley, T., Transformation towards a circular economy in the Australian construction and demolition waste management system. *Sustainable Production and Consumption,* 2022. 30: p. 89-106.

[32] Udawatta, N., Zuo, J., Chiveralls, K. and Zillante, G., Improving waste management in construction projects: An Australian study. *Resources, Conservation and Recycling,* 2015. 101: p. 73-83.

[33] Park, J. and Tucker, R., Overcoming barriers to the reuse of construction waste material in Australia: A review of the literature. *International Journal of Construction Management,* 2017. 17(3): p. 228-237.

[34] Shooshtarian, S., Caldera, S., Maqsood, T., Ryley, T. and Khalfan, M., An investigation into challenges and opportunities in the Australian construction and demolition waste management system. *Engineering, Construction and Architectural Management,* 2022. 29(10): p. 4313-4330.

[35] Shooshtarian, S., Le, T., Feng, Y. and Bettini, L., Analysis of sustainable procurement in supplying recycled content: A case study in Western Australia. *IOP Conference Series: Earth and Environmental Science*, 2022. 1101(4): p. 042004.

[36] SBEnrc P.175. *Project 1.75 creation and stimulation of end markets for construction and demolition waste.* 2021 [cited 2022 11/2021]; Available from: https://bit.ly/3K6tvKF.

[37] Hyder Consulting, Encycle Consulting and Sustainable Resource Solutions, Construction and Demolition Waste Status Report; *Management of Construction and Demolition Waste in Australia.* 2011.

[38] Davis, P., Simon, L., Sher, W., Tang, P. and Newaz, T., Key Solutions For Construction and Demolition (C&D) Waste Management in NSW, Australia, *in AUBEA*. 2019: Noosa, Australia.

[39] *Australian Government (Department of Agriculture WatE),* Sustainable procurement guide: A practical guide for commonwealth entities 2020.

[40] Australian ISO 20400 *Committee, Building the business case for sustainable procurement in Australia – guidance.* 2018.

[41] Hasselbalch, J. A., Costa, N. and Blecken, A., *Investigating the Barriers to Sustainable Procurement in the United Nations,* ed. M. Klumpp, et al., 2015: Springer Cham.

[42] *Australian Government, Sustainable procurement guide: A practical guide for commonwealth entities, Agriculture Water and the Environment,* Editor. 2020, November. CC BY 4.0.: Canberra, Australia

[43] *Australian ISO 20400 Committee, Building the business case for sustainable procurement in Australia – guidance*. 2018: Sydney, Australia.

[44] NSW, L. G., *Sustainable Procurement Guide: for local government in NSW.* 2021.

[45] Costa, B. B. F. D. and Motta, A. L. T. S. D., Key factors hindering sustainable procurement in the Brazilian public sector: A Delphi study. *International Journal of Sustainable Development and Planning,* 2019. 14(2): p. 152-171.

[46] Delmonico, D., Jabbour, C. J. C., Pereira, S. C. F., de Sousa Jabbour, A. B. L., Renwick, D. W. S. and Thomé, A. M. T., Unveiling barriers to sustainable public procurement in emerging economies: Evidence from a leading sustainable supply chain initiative in Latin America. *Resources, Conservation and Recycling*, 2018. 134: p. 70-79.

[47] Bougrain, F., Impact of Green Public procurement on the market of recycled concrete, in *HISER International Conference.* 2017: DELFT (The Netherlands).

[48] McMurray, A. J., Islam, M. M., Siwar, C. and Fien, J., Sustainable procurement in Malaysian organizations: Practices, barriers and opportunities. *Journal of Purchasing and Supply Management*, 2014. 20(3): p. 195-207.

[49] Islam, M. M., Murad, M. W., McMurray, A. J. and Abalala, T. S., Aspects of sustainable procurement practices by public and private organisations in Saudi Arabia: an empirical study. *International Journal of Sustainable Development & World Ecology,* 2017. 24(4): p. 289-303.

[50] Adell, A. and Schaefer, B., *Green Public Procurement in the Republic of Korea: A Decade of Progress and Lessons Learned*. 2019.

[51] UN Environmental Programme, *Sustainable public procurement*. 2016.

[52] Shooshtarian, S., Le, T. T., Feng, Y. and Bettini, L., Analysis of sustainable procurement in supplying recycled content: A case study in Western Australia, in *IOP Conference Series: Earth and Environmental Science*. 2022.

[53] Zhu, Q., Geng, Y. and Sarkis, J., Motivating green public procurement in China: An individual level perspective. *Journal of Environmental Management,* 2013. 126: p. 85-95.

[54] Bohari, A. A. M., Skitmore, M., Xia, B. and Teo, M., Green oriented procurement for building projects: Preliminary findings from Malaysia. *Journal of Cleaner Production,* 2017. 148: p. 690-700.

[55] Sajjad, A., Eweje, G. and Tappin, D., Managerial perspectives on drivers for and barriers to sustainable supply chain management implementation: Evidence from New Zealand. *Business Strategy and the Environment*, 2020. 29(2): p. 592-604.

[56] Ogunsanya, O. A., Aigbavboa, C. O., Thwala, D. W. and Edwards, D. J., Barriers to sustainable procurement in the Nigerian construction industry: an exploratory factor analysis. *International Journal of Construction Management,* 2022. 22(5): p. 861-872.

[57] Pretoria, F.i.s.p.w.t.p.a.p.s.i.S.A.U.o., Factors influencing sustainable procurement within the private and public sector in South Africa University of Pretoria. *2015, Factors influencing sustainable procurement within the private and public sector in South Africa University of Pretoria.*

[58] Shooshtarian, S. and Maqsood, T. *How does product certification help construction and demolition waste resource recovery?* 2021; Available from: https://tinyurl.com/mr7wdc7c.

[59] WSP Environmental and TRL Ltd, Evaluating options for declaring recycled content in construction products and projects. 2005, *The Waste & Resources Action Programme Banbury,* UK.

[60] Silva, R. V., de Brito, J. and Dhir, R. K., Availability and processing of recycled aggregates within the construction and demolition supply chain: A review. *Journal of Cleaner Production,* 2017. 143: p. 598-614.

[61] Equilibrium, *Review of standards and specifications for recycled content products.* 2019: Melbourne, Australia, 20.

[62] Ghaffar, S. 2019; Available from: *https://theconversation.com/how-we-can-recycle-more-buildings-126563*.

[63] Bao, Z. and Lu, W., Developing efficient circularity for construction and demolition waste management in fast emerging economies: Lessons learned from Shenzhen, China. *Science of The Total Environment*, 2020. 724: p. 138264.

[64] Bruvoll, A., Taxing virgin materials: an approach to waste problems. *Resources, Conservation and Recycling*, 1998. 22(1): p. 15-29.

[65] UK's HM Revenue & Customs, *Aggregate Levy Guidance, U.G. Her Majesty's Revenue and Customs, Editor.* 2018: London, UK.

[66] *European Commission, Proposal for a council decision on the system of own resources of the European.* 2018.

[67] Anderson, L. and Gbor, N., *Plastic waste in Australia and the recycling greenwash 2024.*

[68] Shooshtarian, S., Maqsood, T., Khalfan, M., Yang, R. J. and Wong, P., Landfill Levy Imposition on Construction and Demolition Waste: Australian Stakeholders' Perceptions. *Sustainability,* 2020. 12(11).

[69] Shooshtarian, S., Hosseini, M. R., Kocaturk, T., Arnel, T. and T. Garofano, N., Circular economy in the Australian AEC industry: investigation of barriers and enablers. *Building Research & Information*, 2023. 51(1): p. 56-68.

[70] WA Department of Local Government, S.a.C.I. *Local governments rewarded for using recycled construction and demolition waste.* 2022; Available from: https://shorturl.at/3qBeK.

[71] Shooshtarian, S., Caldera, S., Maqsood, T., Ryley, T., Wong, P. S. P. and Zaman, A., Analysis of factors influencing the creation and stimulation of the Australian market for recycled construction and demolition waste products. *Sustainable Production and Consumption,* 2022. 34: p. 163-176.

[72] Hu, Q., Peng, Y., Guo, C., Cai, D. and Su, P. Dynamic Incentive Mechanism Design for Recycling Construction and Demolition Waste under Dual Information Asymmetry. *Sustainability,* 2019. 11, doi: 10.3390/su11102943.

[73] Shooshtarian, S., Maqsood, T., Wong, P. S. P., Caldera, S., Ryley, T., Zaman, A. and Cáceres Ruiz, A. M., Circular economy in action: the application of products with recycled content in construction projects – a multiple case study approach. *Smart and Sustainable Built Environment,* 2024. 13(2): p. 370-394.

[74] Calvo, N., Varela-Candamio, L. and Novo-Corti, I. A Dynamic Model for Construction and Demolition (C&D) Waste Management in Spain: Driving Policies Based on Economic Incentives and Tax Penalties. *Sustainability,* 2014. 6, 416-435 doi: 10.3390/su6010416.

[75] IPWEA, *Specification for Supply of Recycled Material for Pavements, Earthworks and Drainage* 2010. 2010.

[76] *Equilibrium,* Appendix. 2019.

[77] ISC, *IS Rating Scheme*. 2022.

[78] UDIA, *EnviroDevelopment: Developments for environmentally conscious home buyers.* 2023.

[79] International Living Future Institute, *Living Building Challenge*. 2024.

[80] Shooshtarian, S., Maqsood, T., Wong, P. S. and Bettini, L., Application of Sustainable Procurement Policy to Improve the Circularity of Construction and Demolition Waste Resources in Australia. *Materials Circular Economy*, 2022. 4(27): p. 1-22.

[81] NCC, *Building for 7 Stars: top tips and guidance.* 2022.

[82] Code, N. C., *Volume* 2. 2023.

[83] *Sustainability Victoria,* 7 Star Homes Program: Guidelines. 2024.

[84] Damptey, E. O., *Optimising the Use of Recycled C&D Waste Material in Civil Construction Projects*. 2011, Swinburne University: Australia.

[85] Rosen, H. S. and Gayer, T., *Public Finance.* 2008, New York: McGraw-Hill Irwin.

[86] Lu, Y., Zhu, X. and Cui, Q., Effectiveness and equity implications of carbon policies in the United States construction industry. *Building and Environment,* 2012. 49: p. 259-269.

[87] Stavins, R. N., *Experience with Market-Based Environmental Policy Instruments* 2001: Washington, D.C.

[88] *Centre for Climate and Energy Solutions.* Cap and Trade: Key Terms Glossary. Climate Change 101. n.d.

[89] European Comission. *EU Emissions Trading System (EU ETS).* 2005; Available from: https://bit.ly/3tBcsca

[90] N. A. *Carbon taxes.* 2019.

[91] IPART. *Greenhouse Gas Reduction Scheme.* 2003; Available from: https://bit.ly/4beZ2cE.

[92] IPART. *NSW Greenhouse Gas Reduction Scheme - Strengths weaknesses and lessons learned - Final Report* - July 2013. 2013; Available from: https://bit.ly/4beZ2cE.
[93] Hyder Consulting, *Construction and Demolition Waste Status Report.* 2012.
[94] Adams, P. D., Parmenter, B. R. and Verikios, G., An Emissions Trading Scheme for Australia: National and Regional Impacts. *Economic Record,* 2014. 90(290): p. 316-344.
[95] Kumarasiri, J., Jubb, C. and Houghton, K. A. *Direct Action not as motivating as carbon tax say some of Australia's biggest emitters.* 2016; Available from: https://theconversation.com/direct-action-not-as-motivating-as-carbon-tax-say-some-of-australias-biggest-emitters-64562.
[96] Nong, D., Meng, S. and Siriwardana, M. G. *A dynamic evaluation of the impacts of an Emissions Trading Scheme on the Australian economy and emissions levels.* 2016.
[97] Fini, A. A. F. and Forsythe, P., Barriers to reusing and recycling office fit-out: an exploratory analysis of demolition processes and product features. *Construction Economics and Building,* 2020. 20(4).
[98] Chileshe, N., Jayasinghe, R. S. and Rameezdeen, R., Information flow-centric approach for reverse logistics supply chains. *Automation in Construction,* 2019. 106: p. 102858.
[99] Udawatta, N., Zuo, J., Chiveralls, K. and Zillante, G., Attitudinal and behavioural approaches to improving waste management on construction projects in Australia: benefits and limitations. *International Journal of Construction Management,* 2015. 15(2): p. 137-147.
[100] Newaz, M. T., Davis, P., Sher, W. and Simon, L., Factors affecting construction waste management streams in Australia. *International Journal of Construction Management,* 2020. 22(13): p. 2625-2633.
[101] Tam, V. W., Kotrayothar, D. and Loo, Y. C., On the prevailing construction waste recycling practices: a South East Queensland study. *Waste Manag Res*, 2009. 27(2): p. 167-74.
[102] Ratnasabapathy, S., Perera, S. and Alashwal, A., Assessment of waste generation and diversion rates in residential construction projects in Australia, in *CIB World Building Congress 2019.* 2019: Hong Kong SAR, China.
[103] Shooshtarian, S., Maqsood, T., Khalfan, M., Yang, J. R. and Wong, S. P. P., *Landfill levy imposition on construction and demolition waste: Australian stakeholders' perceptions Sustainability*, 2020. 12(11): p. 4496.
[104] Shooshtarian, S., Maqsood, T., Wong, P. S. P., Yang, R. J. and Khalfan, M., Review of waste strategy documents in Australia: analysis of strategies for construction and demolition waste. *International Journal of Environmental Technology and Management,* 2020. 23: p. 1-21.
[105] Jayasinghe, R. S., Rameezdeen, R. and Chileshe, N., Waste management practices in Australia: Comparison of strategies, in *22nd International Conference on Advancement of Construction Management and Real Estate.* 2018: Melbourne, Australia.

[106] Zhao, X., Webber, R., Kalutara, P., Browne, W. and Pienaar, J., Construction and demolition waste management in Australia: A mini-review. *Waste Management & Research,* 2021. 40(1): p. 34-46.

[107] Rameezdeen, R., Chileshe, N., Hosseini, M. R. and Lehmann, S., A qualitative examination of major barriers in implementation of reverse logistics within the South Australian construction sector. *International Journal of Construction Management*, 2016. 16(3): p. 185-196.

[108] Forghani, R., Sher, W., Kanjanabootra, S. and Totoev, Y. Consequence of waste levy revocation: case study Queensland, Australia. in *23rd Annual Pacific Rim Real Estate Society Conference on The Built Environment and Health, Infrastructure and System Resilience Planning*. Sydney, Australia.

[109] Wu, H., Zuo, J., Yuan, H., Zillante, G. and Wang, J., Cross-regional mobility of construction and demolition waste in Australia: An exploratory study. *Resources, Conservation and Recycling*, 2020. 156: p. 104710.

[110] Shooshtarian, S., Maqsood, T., Wong, P. S. P., Khalfan, M. M. A. and Yang, R. J., Review of energy recovery from construction and demolition waste in Australia. *Journal of Construction Engineering, Management & Innovation*, 2019.

[111] Ratnasabapathy, S., Perera, S. and Alashwal, A., A review of construction waste data and reporting systems used in Australia, in *AUBEA Conference* 2019. 2019: Queensland, Australia.

[112] Tam, V. W. Y. and Lu, W., Construction waste management profiles, practices, and performance: a cross-jurisdictional analysis in four countries. *Sustainability,* 2016. 8(2): p. 190.

[113] Xu, J., Ye, M., Lu, W., Bao, Z. and Webster, C., A four-quadrant conceptual framework for analyzing extended producer responsibility in offshore prefabrication construction. *Journal of cleaner production,* 2021. 282: p. 124540.

[114] Thorpe, B. and Kruszewska, I., Strategies to Promote Clean Production: Extended Producer Responsibility. 1999: *Institute for Sustainable Futures*, UTS.

[115] Widmer, R., Oswald-Krapf, H., Sinha-Khetriwal, D., Schnellmann, M. and Böni, H., Global perspectives on e-waste. *Environmental impact assessment review,* 2005. 25(5): p. 436-458.

[116] Nahman, A., Extended producer responsibility for packaging waste in South Africa: Current approaches and lessons learned. *Resources, Conservation and Recycling,* 2010. 54(3): p. 155-162.

[117] Forslind, K. H., Implementing extended producer responsibility: the case of Sweden's car scrapping scheme. *Journal of Cleaner Production*, 2005. 13(6): p. 619-629.

[118] Langrová, V. *Comparative analysis of EPR programmes for small consumer batteries: case study of the Netherlands, Switzerland and Sweden*. 2002.

[119] Lindhqvist, T., Extended Producer Responsibility in Cleaner Production: Policy Principle to Promote Environmental Improvements of Product Systems, *in The International Institute for Industrial Environmental Economics]*. 2000, Lund University.

[120] Gupt, Y. and Sahay, S., Review of extended producer responsibility: A case study approach. *Waste Manag Res,* 2015. 33(7): p. 595-611.

[121] Isenhour, C., Wagner, T., Blackmer, T., Silka, L., Peckenham, J., Hart, D. D. and McRae, J., Moving Up the Waste Hierarchy in Maine: Learning from Best Practice State-Level Policy for Waste Reduction and Recovery. *Maine Policy Review,* 2016. 25(1): p. 15-29.

[122] Srour, I., Chong, W. K. and Zhang, F., Sustainable recycling approach: an understanding of designers' and contractors' recycling responsibilities throughout the life cycle of buildings in two US cities. *Sustainable Development,* 2012. 20(5): p. 350-360.

[123] Shanoff B, Proposed recycling rules create obstacles. *World Wastes,* 1996. 39(6): p. 14-17.

[124] Shooshtarian, S., Maqsood, T., Wong, P. S., Khalfan, M. and Yang, R. J. Extended Producer Responsibility in *the Australian Construction Industry. Sustainability,* 2021. 13, doi: 10.3390/su13020620.

[125] Shooshtarian, Salman, Maqsood, T., Wong, P. S., Khalfan, M. and Yang, R. J., Extended producer responsibility in *the Australian construction industry. Sustainability,* 2021. 13(2): p. 620.

[126] Ratnasabapathy, S., Alashwal, A. and Perera, S., Exploring the barriers for implementing waste trading practices in the construction industry in Australia. *Built Environment Project and Asset Management,* 2021. 11(4): p. 559-576.

[127] Shooshtarian, S., Maqsood, T., Barrett, C., Wong, P. S. P., Yang, R. J. and Khalfan, M., Opportunities to reduce brick waste disposal Imaginable Futures: Design Thinking, and the Scientific Method, in *54th International Conference of the Architectural Science Association 2020*. 2020: Virtual Conference: Auckland, New Zealand.

[128] Zuo, J. and Zhao, Z. Y., Green building research–current status and future agenda: A review. *Renewable and Sustainable Energy Reviews*, 2014. 30: p. 271-281.

[129] Kabirifar, K., Mojtahedi, M. and Wang, C. C., A systematic review of construction and demolition waste management in Australia: Current practices and challenges. *Recycling,* 2021. 6(2): p. 34.

[130] Laviano, H., Jordan, B., Monica, T. and Noor, D., *Waste and recycling industry in Australia: A Submission to the Senate Inquiry.* 2017.

[131] Glanville, K. and Chang, H. C., Mapping illegal domestic waste disposal potential to support waste management efforts in Queensland, Australia. *International Journal of Geographical Information Science,* 2015. 29(6): p. 1042-1058.

[132] Ratnasabapathy, S., Perera, S. and Alashwal, A., A review of smart technology usage in construction and demolition waste management, in *Proceedings of the 8th World Construction Symposium*. 2019: Colombo, Sri Lanka.

[133] Queensland Government, *Waste–Everyone's responsibility*. 2014.

[134] Dawkins, E. and Allan, P., Landfill ban investigation. *Australian Government.* 2010.

[135] Shooshtarian, S., Maqsood, T., Zaman, A., Caldera, S., Ryley, T. and Wong, P. S., Enhancing the Use of Products with Recycled Content in the Building and Construction Sector. *Waste and Waste Management.* 2024: Nova science publishers. 238.

[136] Murray, R. L., *Construction and demolition waste in Western Australia: a case study on best practice demolition.* 2019, Murdoch University.

[137] Bao, Z., Lee, W. M. W. and Lu, W., Implementing on-site construction waste recycling in Hong Kong: Barriers and facilitators. *Science of The Total Environment,* 2020. 747: p. 141091.

[138] Chen, Z., et al., Construction, renovation, and demolition waste in landfill: a review of waste characteristics, environmental impacts, and mitigation measures. *Environmental Science and Pollution Research,* 2022. 29(31): p. 46509-46526.

[139] Liu, J., Teng, Y., Jiang, Y. and Gong, E., A cost compensation model for construction and demolition waste disposal in South China. *Environmental Science and Pollution Research,* 2019. 26: p. 13773-13784.

[140] WMRR, Strategic Business Plan 2023-2026. *2023, Waste Management and Resource Recovery Association of Australia: Sydney, Australia.*

[141] Government, A. *Waste exports.* 2021; Available from: https://www.dcceew.gov.au/environment/protection/waste/exports.

[142] Shooshtarian, S., Maqsood, T., Wong, P. S., Khalfan, M. and Yang, R. J., *The impact of new international waste policies on the Australian construction and demolition waste stream, in AUBEA 2021: Construction Education – Live the Future.* 2021: Virtual. p. 635-644.

[143] ACOR, *Waste and Recycling Industry Fire Forum Report.* 2022.

[144] *Australian Government,* Product stewardship in Australia. 2023.

[145] *Australian Government,* Australian Packaging Covenant Organisation (APCO) Arrangement. 2023.

[146] *Australian Government,* Product stewardship accreditation. 2023.

[147] *Product Stewardship Centre of Excellence.* 2021; Available from: https://stewardshipexcellence.com.au/.

[148] *Australian Government, Minister's product stewardship priority list.* 2023.

[149] Product Stewardship Centre of Excellence. *Evaluating product stewardship benefits and effectiveness.* 2023; Available from: https://shorturl.at/cwGV7.

[150] Low, J. K., Wallis, S. L., Hernandez, G., Cerqueira, I., Steinhorn, G. and Berry, T. A., Encouraging circular waste economies for the New Zealand construction industry: Opportunities and barriers. *Frontiers in Sustainable Cities,* 2020. 2: p. 35.

[151] EPA Victoria, *Impact of Landfill levy on the steel recycling sector in Victoria.* 2007: Melbourne, Australia.

Chapter 8

Insights from a Survey on Circular Economy Policies

Salman Shooshtarian[1]
Muzna Anam[1]
Savindi Caldera[2]
and Tayyab Maqsood[1]
[1]School of Property, Construction and Project Management, RMIT University, Melbourne, Australia
[2] School of Science, Technology & Engineering, University of the Sunshine Coast, Petrie, Australia

Abstract

This chapter provides insights into the current state of circular economy (CE) policies for the integration of products with recycled content (PwRC) within the building and construction sector. Through a survey study, the study captures the perspectives of key stakeholders on 17 CE policies. The findings of the study outline the stakeholders' familiarity both, in general and with surveyed CE policies. It explores the effectiveness and impacts, positive and negative along with a comparison of both impacts of these policies on the optimal uptake of PwRC. It also highlights the stakeholder's identified challenges, perspective for adoption at the national level and the proposed strategies to enhance the adoption of PwRC in the sector. The analysis of findings shows that key policies driving the adoption of PwRC, are Extended Producer Responsibility (EPR), Sustainable Procurement, Environmental Sustainability Rating, and the Carbon Tax Scheme emerging as the most influential across multiple impact mechanisms. These policies play a crucial role in building end-user confidence, creating demand, ensuring

In: Policy, Practice and Education of the Circular Economy in Built Environments
Editors: Salman Shooshtarian, Peter SP Wong and Tayyab Maqsood
ISBN: 979-8-89530-623-9

sustainable supply chains, improving quality, reducing costs, and enhancing environmental and social benefits in construction. The chapter provides policymakers, industry leaders, and peak bodies with a clear understanding of actionable pathways to enhance CE policy efficacy for enhancing the use of PwRC in the sector.

1. Introduction

The adoption of products with recycled content (PwRC), in alignment with CE objectives, offers a pathway to reduce the environmental and economic costs of waste, including landfill maintenance, virgin material extraction, landfill levies, transportation expenses and illegal dumping [1].

Given the current satisfactory waste recovery rates in Australia, the next critical challenge is maximising the use of PwRC in construction projects. The uptake of these resources is reported to be influenced by three key factors: education, enforcement and encouragement - all of which can be effectively promoted through the development and enforcement of appropriate CE-based policies. Hence, CE policies are essential in promoting and guiding the optimal integration of these resources within the sector.

Researchers agree that "*the transition to a circular built environment is key to achieving a resource-effective and sustainable society*" [2, p. 636]. However, according to Hossain, Ng [3], there seems to be a lack of policy interventions to support the transition to CE [3]. Both Hossain, Ng [3] and Osobajo, Oke [4] conducted literature reviews on CE in the building and construction sector, and both discovered that there is scant research on CE policies that could support this transition. Thus, further research is essential to determine which policy measures can facilitate the development of a circular economy in the sector.

Although some CE policies exist within the Australian context, there remains limited insight into their effectiveness in stimulating demand for PwRC [5]. Gaining a deeper understanding of these policies' performance is essential for informing the development of new policies and refining the current policy framework. This knowledge can help ensure that policies are not only present but are also impactful in promoting a strong CE. Stakeholders' perceptions significantly influence the use of PwRC in construction projects [6]. However, it is not yet clear how key stakeholders in this sector perceive the use of PwRC in construction projects. Therefore, this study aims to understand how the sector's key stakeholders perceive the relevance, significance and impact of the policy approach in optimising the use of PwRC in construction projects using a survey from Australian stakeholders. The rationale for this research is the critical need to inform, empower and engage key stakeholders to advance the agenda for increased utilisation of PwRC in both government and industry sectors.

2. Research Objectives

The research aimed to understand the perceptions of key stakeholders in the Australian building and construction sector regarding CE policies that impact the optimal use of PwRC within the industry. To achieve this aim, the following objectives were addressed:

1. To explore stakeholders' familiarity with these policies within the sector
2. To evaluate the effectiveness and impact of these policies on the optimal uptake of PwRC in the sector
3. To assess the challenges and implementation strategies associated with these policies.

3. Research Approach

This section of the chapter outlines the research aim and objectives of this study. It then details the approach used to capture the knowledge and perspectives of key Australian stakeholders on the identified 17 policies that directly or indirectly influence the optimal utilisation of PwRC in the building and construction sector. These policies are identified and explored in Chapter 7. Lastly, it presents profiles of the research participants, including their demographic information.

3.1. Research Design

The research utilised a mixed-methods approach to assess the impact of key CE policies on the adoption of PwRC in the building and construction sector. To address Objectives 1 through 3, quantitative data was collected via a semi-structured survey. Surveys are an appropriate data collection method to obtain quantitative and qualitative information, for well-planned questionnaires are widely used by researchers within the CE and policy analysis domains [7-9]. This study used an online questionnaire as the most appropriate delivery method because this tool is efficient and flexible and ensures participant confidentiality. In addition, conducting questionnaires online is the most common delivery method, so participants are familiar with the approach and are more likely to respond [10].

3.1.1. Sample and Data Collection

A cross-sectional survey of a purposive sample of stakeholders of C&D waste management operating in different jurisdictions of Australia was conducted from January to September 2024. A purposive sampling strategy was the most time-effective sampling approach available. It was employed to recruit a wide range of participants across the C&D waste supply chain. Recruitment was executed according to the Australian National Statement on Ethical Conduct in Human Research and RMIT University Human Ethics Committee instructions and requirements. Several key industry peak bodies and public organisations assisted in recruiting participants for the survey.

Since the study required experts in the field, one of the main selection criteria was adequate experience in dealing with the management of C&D waste and CE in Australia. Participants were recruited through email communication, LinkedIn posts and direct messaging. An email or LinkedIn message containing the online survey link and the project information sheet was sent to a list of participants compiled by the research team in a single round. A reminder email/message was also sent to those who did not respond in the first round. Survey participation was voluntary, and a completed survey implied informed consent. The investigators maintained the privacy and confidentiality of all survey information as per the human ethics requirements.

3.1.2. Survey Design

The questionnaire survey was developed based on the review of the literature, regulations, policy statements, other international guidelines and expert opinions regarding the main issues of PwRC uptake and implementation of CE policies. The questionnaire was designed using Qualtrics. The first draft of the questionnaire was reviewed by three experts to establish content validity and tested for online accessibility and comprehension by a group of construction professionals and experienced researchers. The content, clarity and length of the survey were modified accordingly. The questionnaire included multiple questions for 17 identified policies, addressing participants' familiarity with it, their perceptions of its effectiveness, any unintended negative outcomes, suggestions for improvement, and the scope of its applicability. The survey design consisted of a range of question types: multiple-choice questions, 5- and 7-point Likert scales (1 = 'strongly agree' to 7= 'strongly disagree'), and single and multiple text entries. The survey consisted of open- and close-ended questions. Close-ended survey questions offer a set of answer alternatives for participants to choose from [11]. Open-ended questions allow researchers to hear directly from both public and private sectors how current policies can either help or hinder the successful application of PwRC in construction projects, revealing the reasons behind these views.

3.1.3. Research Participants' Profiles

The administration of the survey resulted in 92 responses, of which 62 were complete and therefore were included in the data analysis. Figure 1 presents a summary of the survey participants' profiles. Most respondents were in the age groups of 35-44 and 45-54 (61%), and almost half of this population had been promoting material circularity within the sector for between 1 and 5 years (35.5%). Their main locations of operation were in four major Australian states, with Victoria having the greatest number of respondents (40.3%). Figure 2 presents the participants' employment sector. The main employer sectors among the survey participants were 'Government' (25.8%) followed by 'Consultation', 'Construction' and 'Academia' (14.5%). Participants from academia were those recently engaged in research on C&D waste management and implementing CE principles within the built environment. The sub-sectors of the government category included general, infrastructure delivery, policy-making and enforcement. The 'Material supply' category refers to those engaged in material engineering, manufacturing and supplying.

Stakeholder perspectives are crucial in policymaking as they bring diverse knowledge and values, leading to more informed and effective policies [12, 13]. Engaging stakeholders ensures relevant information and public values are considered, promoting ownership, responsibility, and policy acceptance [14]. This minimises disputes, increases collaborative problem-solving, and results in inclusive, sustainable solutions. In C&D waste management, stakeholder engagement is essential for transitioning to a CE by understanding complex material flows and developing effective strategies [12].

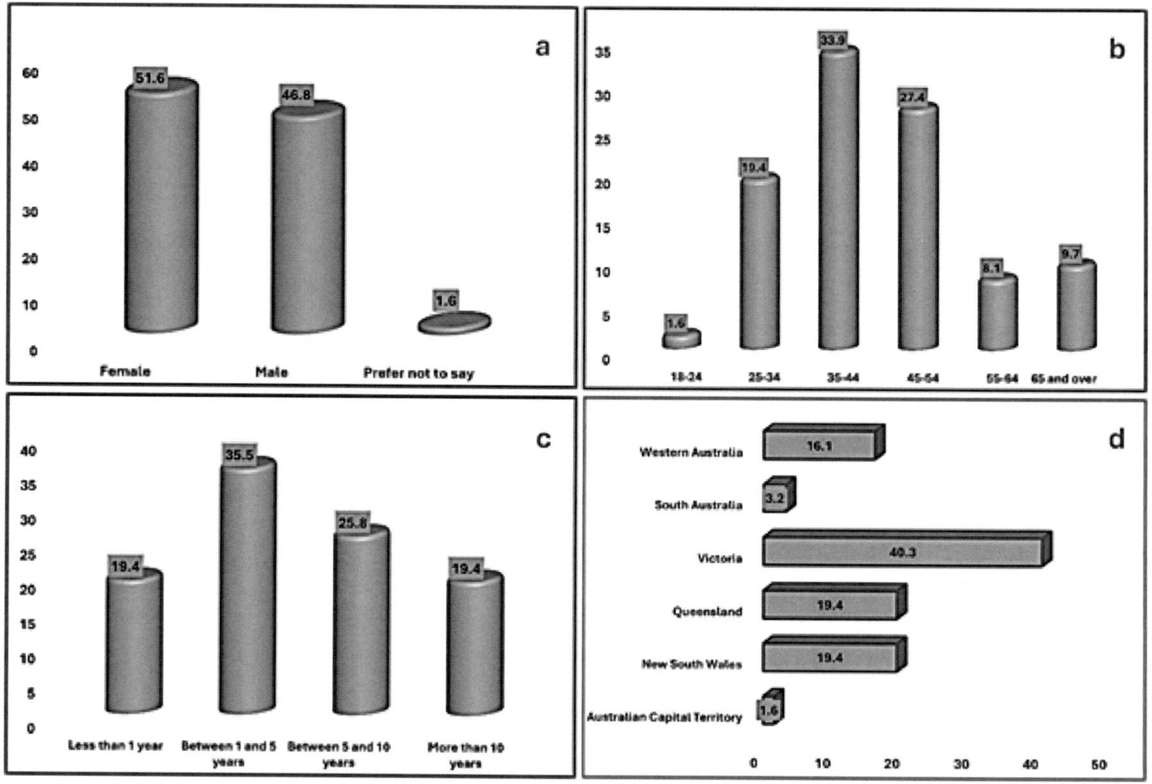

Source: Authors.

Figure 2. Participants profile (frequency distribution %). Gender (a), age groups (b), work experience (c) and location of operation (d).

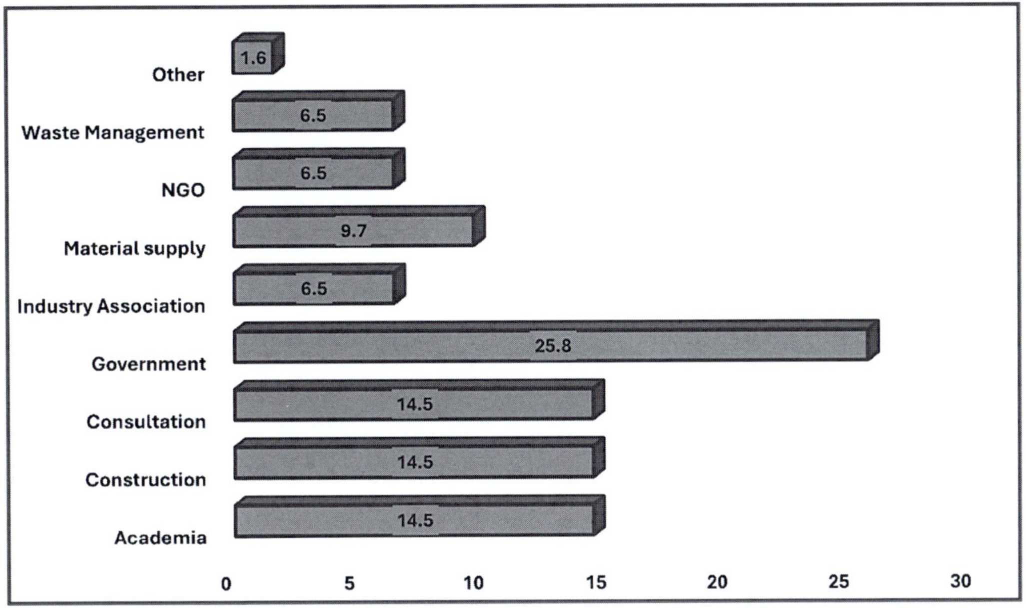

Source: Authors.

Figure 3. Employment sector of participants (frequency distribution %).

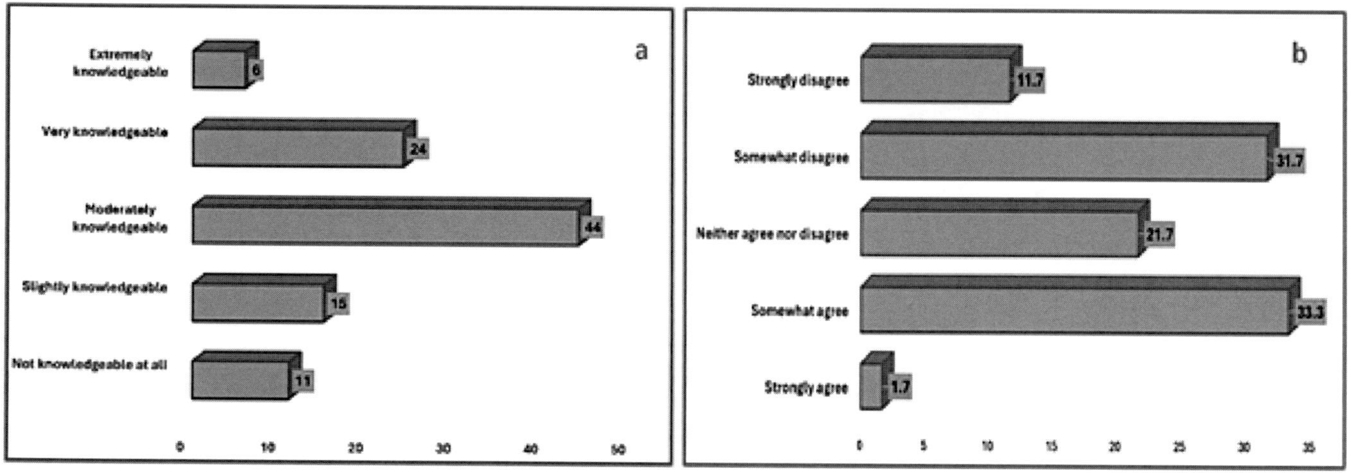

Source: Authors.

Figure 3. Frequency distribution (%) of participants' responses on (a) their level of general knowledge of CE policies and (b) perceptions on whether existing Australian policies support use of PwRC in construction.

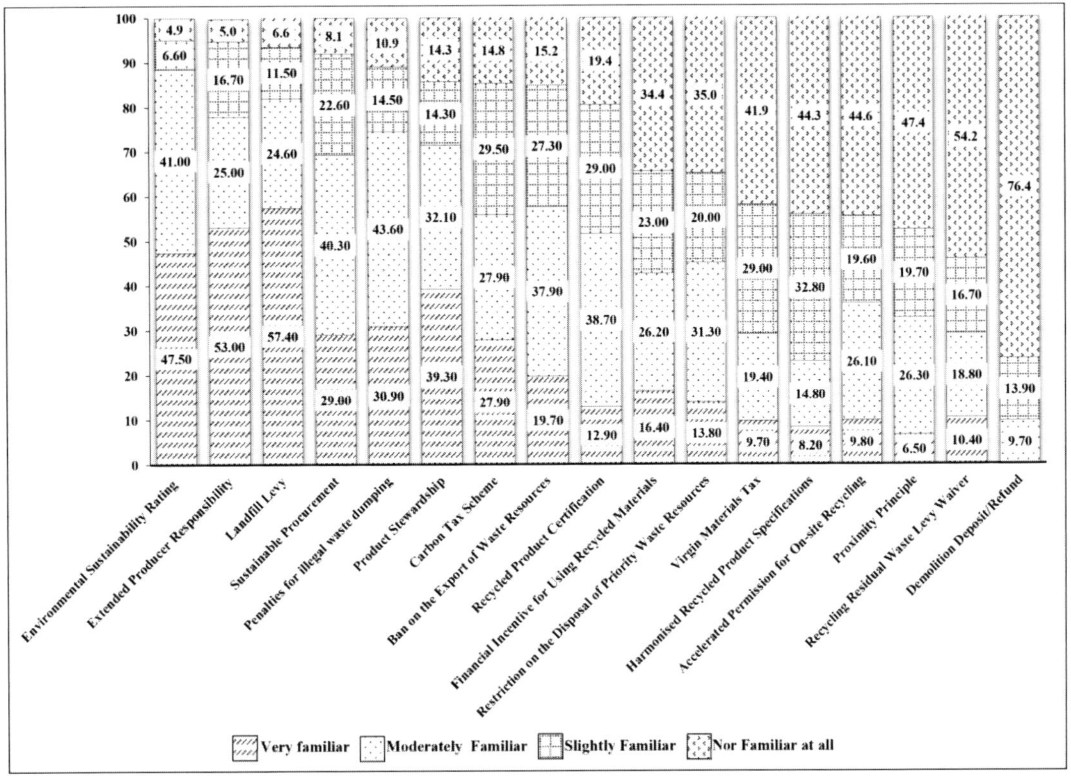

Source: Authors.

Figure 4. Comparative overview of familiarity levels across study policies (frequency distribution %).

4. Findings

The findings are presented and discussed in three themes, each addressing the objectives of the study: (1) the familiarity of participants regarding CE policies, both as general knowledge and specific to policies (2) Responses to CE policies effectiveness along with negative and positive impacts and lessons learned; (3) the challenges for implementation strategies and comparative analysis of responses for further feedback.

4.1. Familiarity with Policy in the Sector

This section addresses objective 1 of this study by understanding the stakeholder's familiarity and understating of the CE policies in general and specifically for the identified policies.

4.1.1. General Knowledge of Circular Economy Policies

Based on the 17 policies presented in the previous chapter, the participants were asked to self-assess their knowledge of Australian CE policies that influence the use of PwRC in the construction sector. Nearly two-thirds indicated a moderate to extensive understanding of these policies (Figure 3a). When asked whether existing Australian policies actively support the optimal use of PwRC, only 35% of participants agreed that these policies effectively serve this purpose. This suggests that the majority view current policy frameworks as insufficiently aligned with promoting the use of recycled content in construction (Figure 3b).

4.1.2. Levels of Understanding of Surveyed CE Policy

To determine which policy is most familiar to research participants, a comparative analysis was conducted. As shown in Figure 4, the level of participants' understanding of each policy varies significantly, ranging from 9.7% for the demolition deposit-refund policy (moderately familiar) to 78.3% for the environmental sustainability rating policy (very familiar to moderately familiar). Consequently, the least known policies - where fewer than one-fourth of participants reported moderate to high familiarity - included the demolition deposit refund, waiver for recycling residual waste, proximity principle, and harmonised PwRC specifications. The results also show that out of the 16 policies, only 5 were properly (very familiar and moderately familiar) known to 60% or more of the participants. The most known circular economy policies included *Environmental Sustainability Rating* (78%), *Penalty on Illegal Waste Dumping* (74.5%), *Landfill Levy (Tax)* (82%) and *Product Stewardship* (71.4%). The four policies that are least known (not familiar) to participants include '*Demolition Deposit/Refund*' (76.4%), '*Recycling Residual Waste Levy Waiver*' (54.2%), '*Virgin Materials Tax*' (41.9%), and '*Accelerated Permission for On-site Recycling*' (44.6%).

4.2. Policy Effectiveness and Impact

This section addresses objective 2 of this study by outlining the effectiveness of these policies. It also highlights both positive and negative impacts along with the comparison, stated by the stakeholders for the application of each policy. Table 1 shows the details of the effectiveness, impacts and measures for each policy based on the participant's responses.

4.2.1. Policy Effectiveness

The participants were asked to share their views about the effectiveness of the policies surveyed. In this chapter, it is regarded as overall effectiveness (shown in Column 1, Col 1). Generally, the majority of participants indicated that all policies in this study are effective in enabling the use of PwRC in construction projects to various degrees (moderately effective to very effective). As can be seen in Table 1, according to the participants, the three most overall effective (Col 1) policies were *Landfill Tax* (73%), *Environmental Sustainability Rating* (77%), and *Penalty on Illegal Waste Dumping* (68%). Conversely, only for the 'sustainable procurement' (34%) policy a notable portion of participants found its application slightly effective or ineffective (Col 2). Furthermore, a significant number of participants were unsure or did not know (Col 3) about the effectiveness of three policies: *Levy waiver on waste recycling residuals* (50%) and *Demolition Deposit/Refund* (46%).

To better understand the effectiveness of the application of the policies in Australia, further analysis was conducted. This analysis involved a comparison between the overall and actual (current) effectiveness of these policies. In this chapter, actual effectiveness (shown in Col 4) is calculated through the comparison for which the policies that are not currently applied in Australia and the 'not sure/do not know' category responses were excluded from the analysis. This was to ensure that the comparison was based on the participants' existing knowledge of the study policies' effectiveness. Therefore, the four policies at are not currently applied in Australia including '*Tax on Virgin Materials', 'Carbon Tax Scheme', 'Demolition Deposit/Refund' and 'Ban on the Export of C&D Waste Resources'* were excluded, indicated in italics in Table 1. The highest actual effectiveness is shown by two policies, *Environmental Sustainability Rating* (66%) and *Landfill Tax* (66%), which also showed significant overall effectiveness as well. The higher diversion values in Table 1 (Col 6) suggest that the effectiveness of these policies is generally lower than expected. This is more evident in four cases: *Financial Incentive for Procurement of Products* (29), *Proximity Principle* (22%), *Accelerated Permission for On-site Recycling* (21%) and *Harmonised Recycled product certification* (21%). However, no diversion is shown by *Product Stewardship.*

4.2.2. Positive Impacts of Policies

The participants identified the positive impacts of each policy from the provided options in the questionnaire, and this can also act as an impact mechanism for each policy. Table 2 provides the frequency distribution of responses for each policy impact mechanism and the highlighted cell indicates the mechanism that can help the policy to boost PwRC use. Participants were further asked to give their inputs in the form of open-ended questions in the survey, to provide the impacts that can enhance the PwRC use. The responses for positive impacts of each policy are summarised in Table 3.

4.2.3. Negative Impacts of Policies

The participants were asked about the negative impacts and challenges they were aware of when implementing each policy at the time of the survey. The responses for the highest negative impacts were for two policies, '*Demolition Deposit/Refund (79%)'* and *'Levy waiver on waste recycling residuals* (72%)*'*, as shown in Table 1 (Col 6) and the lowest was for *'Penalties for illegal waste dumping* (7%), *'Proximity Principle* (9%) *and 'Accelerated Permission for On-site Recycling* (10%)*'*.

In the survey questionnaire, if participants reported that there were negative consequences associated with the implementation of a policy, they were also required to specify what these consequences were. As shown in Table 1 (Col 7), some policies were reported to have multiple negative effects (e.g. *Sustainable Procurement, Environmental Sustainability Rating, Ban on the Export of C&D Waste Resources*), only a few policies (e.g. *Penalties for illegal waste dumping* and *Demolition Deposit-Refund*) were associated with a small number of negative impacts - possibly due to limited awareness of these policies. A maximum of 14 negative impacts and a minimum of 1 negative impact for a policy have been recorded from the response. The summary of participants' comments about these negative impacts is shown in Table 4-5.

4.2.4. Positive Versus Negative Impacts

The participants were required to report the positive or negative impacts of each policy. Figure 5 provides a comparative overview of the positive versus negative impacts of the study's policies. Participants identified *sustainable procurement* (90%), *financial incentives* (81.7%), *recycled product certification* (81.7%), and *product stewardship* (80%) as the top four policies with significantly more positive than negative impacts, also can be seen in Table 1 (Col 8). Conversely, participants noted that the *proximity principle* (41.7%) and *virgin material tax* (46.7%) are the most challenging policies, potentially hindering the optimal use of PwRC in the building and construction sector.

4.3. Challenges and Implementation Strategies of Policy

This section of chapter addresses the objective three of the study by outlining the challenges and implementation strategies for each policy. It outlines the trends for potential application of these policies at are national level and the key measures that can be taken for each policy to mitigate the negative impacts and facilitate the application.

4.3.1. Scope of Application

The participants were asked about the application of the surveyed policies at the national level and the results are shown in Figure 6. In response to this question, over 65% of participants suggested that certain policies should be implemented at a national level (Figure 6a). Among those advocating for national application, participants identified specific policies they deemed most critical for national implementation. The results indicate that *sustainable procurement* (80%), *recycled product certification* (~71%), *financial incentives* (~66%), and a *carbon tax scheme* (~63%) are prioritised for application on a national scale (Figure 6b).

Table 1. Summary of findings of effectiveness, impacts, implementation and challenges of CE policy responded to by stakeholders (Policies in italics are currently not implemented in Australia and the highlighted cells show the highest contribution in respective factor)

	Factors/CE policy	Col 1	Col 2	Col 3	Col 4	Col 5	Col 6	Col 7	Col 8	Col 9	Col 10
		Effectiveness					Impacts			Implementation	Challenges
		Overall effectiveness			Actual effectiveness	Diversion in overall & actual effectiveness	Negative Impacts		More positive than negative impacts	Nationwide implementation recommendation	No. Of key measures proposed
		Very effective to moderately effective	Slightly effective to not effective	Not sure/Not known	Highly to moderately effective	Highly to moderately effective	Policy's application	Indicated potential negative consequences			
		%	%	%	%	%	%		%	%	
Direct Impacts	Sustainable Procurement	58	34	8	42	16	34	12	90	80	19
	Recycled Product Certification	51	~26	22	31	20	16	6	82	71	28
	Tax on Virgin Materials	63	16	21	27		26	9	47	~43	26
	Financial Incentive for Procurement of Products	64	34	1	35	29	20	8	~82	~66	35
	Harmonised Recycled product certification	62	7	31	41	21	13	4	60	~40	29
	Environmental Sustainability Rating	77	11	11	66	11	33	14	68	~57	29
	Carbon Tax Scheme	59	21	20	28		20	3	~57	~63	29

	Factors/CE policy	Col 1	Col 2	Col 3	Col 4	Col 5	Col 6	Col 7	Col 8	Col 9	Col 10
		Effectiveness					Impacts			Implementation	Challenges
		Overall effectiveness			Actual effectiveness	Diversion in overall & actual effectiveness	Negative Impacts		More positive than negative impacts	Nationwide implementation recommendation	No. Of key measures proposed
		Very effective to moderately effective	Slightly effective to not effective	Not sure/Not known	Highly to moderately effective	Highly to moderately effective	Policy's application	Indicated potential negative consequences			
		%	%	%	%	%	%		%	%	
Indirect Impacts	Landfill tax	73	22	5	66	7	46	9	~63	~51	23
	Extended producer responsibility (EPR)	63	23	13	53	10	45	11	65	60	25
	Penalties for illegal waste dumping	68	28	7	53	15	7	2			18
	Restriction on the Disposal of Priority Waste	61	24	12	45	16	15	5	~55	~37	17
	Accelerated Permission for On-site Recycling	55	24	32	34	21	10	7	65	~23	12
	Demolition Deposit/Refund	40	14	46	10		79	1	~47	~34	18
	Proximity Principle	47	18	34	25	22	9	7	~42	~34	15
	Ban on the Export of C&D Waste Resources	50	21	29	47		47	14	~53	~51	11
	Product Stewardship	59	27	14	59	0	~18	8	80	50	15
	Levy waiver on waste recycling residuals	27	23	50	13	14	72	4	~47	~11	15

Source: Authors.

Table 2. The frequency distribution (n) of responses for the policies' impact mechanisms

	Impact mechanism/CE policy	Build confidence among end users	Create demand for PwRC	Create sustainable supply chains	Improve the environmental sustainability	Improve the quality of PwRC	Improve the social image of end-user organisations	Improve the working conditions	Influence decision-making process	Provide a sustainable supply of PwRC	Reduce the cost of PwRC	Others
Direct Impacts	Sustainable Procurement	21	29	31	27	17	18	3	27	21	18	3
	Recycled Product Certification	21	10	16	20	15	17	4	16	14	6	2
	Tax on Virgin Materials	6	36	26	31	11	14	2	24	18	18	5
	Financial Incentive for Procurement of Products	7	32	27	23	8	15	4	23	19	23	3
	Harmonised Recycled product certification	24	14	21	23	22	13	10	19	12	6	4
	Environmental Sustainability Rating	25	24	19	29	12	25	11	29	15	9	4
	Carbon Tax Scheme	2	29	21	36	15	16	4	28	17	20	12
Indirect Impacts	Landfill tax	5	13	21	34	7	3	3	20	15	10	3
	Extended producer responsibility (EPR)	36	34	33	39	32	28	6	28	35	20	6
	Penalties for illegal waste dumping	3	5	6	15	0	10	4	8	5	3	7
	Restriction on the Disposal of Priority Waste	7	12	22	37	11	10	10	23	24	13	6
	Accelerated Permission for On-site Recycling	9	15	24	38	15	20	11	25	23	25	6
	Demolition Deposit/Refund	6	18	23	31	8	14	9	26	27	17	7
	Proximity Principle	2	9	18	40	3	14	6	22	15	21	10
	Ban on the Export of C&D Waste Resources	7	15	31	25	6	10	6	16	25	13	12
	Product Stewardship	6	26	9	34	12	14	7	20	19	10	5
	Levy waiver on waste recycling residuals	4	5	3	25	3	7	5	15	9	12	8

Source: Authors.

Table 3. Stakeholder's responses to the positive impacts of policies

	Impacts/Policy/Name	Impact 1	Impact 2	Impact 3	Impact 4
Direct Impacts	Sustainable Procurement	Increases sustainable supply and drives demand for PwRC verification, traceability, and improved metrics beyond current LCA capabilities	Supports job creation through increased use and demand for PwRC		
	Recycled Product Certification	Ensures traceability of materials within the supply chain	Potential to improve transparency and standards for PwRC		
	Financial Incentive for Procurement of Products	Increased investment in innovation to support PwRC development			
	Harmonised Recycled product certification	Provides greater confidence within the engineering and process engineering professions			
	Carbon Tax Scheme	Financial benefits for government and listed companies	Implementation of bogus offset schemes		
Indirect Impacts	Landfill tax	Seen as a part of the building price scheme now			
	Extended producer responsibility (EPR)	Implications for the design and preparation of products to enable 2nd and 3rd life cycles	Reduces free riders, improves social and environmental conditions, increases economic development, supports local jobs	Forces suppliers and manufacturers to take responsibility for their products' end-of-life impact	Creates an accurate life cycle of the products
	Penalties for illegal waste dumping	Aims to limit harmful waste products in the environment, e.g., asbestos dumping	Penalises wrong-doers, but only if they are caught		
	Restriction on the Disposal of Priority Waste	Improves investment confidence in recycling infrastructure			

Table 3. (Continued)

	Impacts/Policy/Name	Impact 1	Impact 2	Impact 3	Impact 4
Indirect Impacts	Accelerated Permission for On-site Recycling	Streamlined approvals and on-site recycling incentives lower operational costs, benefiting businesses and the environment.	It encourages innovative recycling concepts specific to construction sites	Streamline processes to increase material availability and improve waste processing efficiency	On-site recycling reduces costs and emissions by eliminating transport and double-handling
	Demolition Deposit/Refund	Incentives like deposit/refund schemes can encourage circular practices in construction, supporting circular economy goals			
	Proximity Principle	Ensures traceability of materials within the supply chain	▪ At best, lowers carbon emissions created by logistics ▪ Reduced carbon emissions	A proximity principle for construction and demolition waste can support a regional recycling market and reduce emissions	Local recycling reduces emissions and supports job creation in the recycling industry, contributing to market development
	Ban on the Export of C&D Waste Resources	Industries like steel and aluminium are still exporting to Asia, which should stop supporting local jobs and industry	Restricting exports reduces emissions from transport, supporting the carbon goals of circular economy policies and favouring domestic processing		
	Product Stewardship	Creates production opportunities for PwRC	Take-back initiatives allow for the recycling of materials at end-of-life, supporting circular economy objectives		

Source: Authors.

Table 4. Stakeholder's responses to the negative impacts of policies with direct impacts

Policies/ Impacts	Sustainable Procurement	Recycled Product Certification	Tax on Virgin Materials	Financial Incentive for Procurement of Products	Harmonised Recycled product certification	Environmental Sustainability Rating	Carbon Tax Scheme
Impact 1	Until Australian standards address PwRC application in construction and wherein the design in the buildings, it can have some adverse effects such as lack of durability and unfavourable mechanical properties	Mixed opinions on the reliability of certifying bodies, potentially diminishing credibility (e.g., Green Tag)	Not all PwRC are suitable replacements for virgin materials in certain applications	Ensuring that genuine PwRC are used, preventing the misuse of "fake" PwRC	Policies may inadvertently limit creative applications of PwRC, restricting innovation	Misalignment of policy application, focusing on construction/design instead of operational performance	Traditional LCA often shows higher GGE for PwRC compared to virgin materials, potentially disadvantaging some PwRC
Impact 2	There are no studies to ensure that there are no long-term structural or environmental issues in using these materials	More administrative processes, potentially slow down project timelines	Rising costs in construction can strain budgets and make recycled options less attractive	Concerns over the quality and durability of some PwRC compared to virgin materials	Excessive bureaucracy and regulatory requirements could stifle innovation over time	Additional expenses incurred to achieve higher star ratings, impacting project budgets	Emphasis on carbon-neutral materials may lead to the use of less sturdy materials unable to withstand extreme conditions
Impact 3	Incorrect and/or inadequate application of the policy for its intended purpose	Additional expenses on builders already facing financial pressure	Effective implementation of the policy requires an adequate PwRC supply; otherwise, a tax could be unfair	Risk of improper policy application by stakeholders aiming for financial gains without achieving true sustainability	High standards and regulations may deter new entrants from participating, limiting diversity in the market	Significant costs associated with training personnel in green star qualifications and maintaining certifications	Risk of greenwashing and reliance on ineffective methods, like carbon offsetting, rather than true sustainable practices
Impact 4	Gives the impression that government is doing more than it actually is. A form of greenwashing	Certification requirements may discourage suppliers due to time and cost burdens	Policies may indirectly place additional financial pressures on citizens	Builders may misuse PwRC in unsuitable applications, potentially compromising safety	Different states have unique legislative approaches and perceptions, complicating unified policy implementation	Practices that create an impression of supporting PwRC without substantial impact	

Table 4. (Continued)

Policies/ Impacts	Sustainable Procurement	Recycled Product Certification	Tax on Virgin Materials	Financial Incentive for Procurement of Products	Harmonised Recycled product certification	Environmental Sustainability Rating	Carbon Tax Scheme
Impact 5	Lack of materials testing has led to some materials being rejected for use	Hinders the development of more economically supportive policies	Incentivising PwRC could lead some builders to misuse them in critical areas, risking structural integrity	Need to maintain a balance between the costs of virgin, PwRC and associated taxes		Policies contributing to higher overall construction costs	
Impact 6	It limits the number of suppliers as not all of them source PwRC	Creates a new hurdle for new entrants in the sustainable materials market	Incentivising PwRC suppliers may drive others to exploit loopholes to continue supplying non- PwRC	Current policies may inadvertently favour heavier materials, potentially unbalancing sustainable material choices		Strict standards may prevent the use of innovative PwRC in construction, necessitating subsidies to offset costs for compliance	
Impact 7	Time-consuming, expensive, with more reliance on imports		Penalising suppliers without a constructive approach may not effectively support the policy's goals	Subsidies provided to producers may not incentivise actual material use unless extended to builders or developers with compliance checks		Lack of adequate enforcement and the presence of loopholes reduce policy effectiveness	
Impact 8	No other actions will take place		In the UK, businesses often find paying a tax more affordable than adapting to PwRC use, reducing policy impact	Absence of research confirming that PwRC use does not result in chemical leaching or the creation of harmful by-products over time		More work is required to incorporate PwRC, adding complexity to construction processes	

Policies/ Impacts	Sustainable Procurement	Recycled Product Certification	Tax on Virgin Materials	Financial Incentive for Procurement of Products	Harmonised Recycled product certification	Environmental Sustainability Rating	Carbon Tax Scheme
Impact 9	I don't think that there will be some negative consequences. However, it may take some time to adopt this policy and receive the positive consequences		Businesses tend to absorb or pass on tax costs if the tax rate isn't substantial enough to incentivise sustainable practices			Initial failures in recycled polymer products highlighted the need for thorough testing and adaptation, now leading to successful application in suitable use cases	
Impact 10	Limited certifications and reporting standards can lead to inconsistent outcomes					Rating systems may prioritise ratings over actual environmental impact, potentially diminishing genuine sustainability efforts	
Impact 11	Setting expectations that are too long reduces the rate of improvement					High standards and complex processes can deter new entrants, limiting competition and innovation in the market	
Impact 12	▪Materials are low quality and may be contaminated ▪Early concerns about contamination of product, particularly with asbestos ▪Ongoing concerns around the performance of some products compared to virgin materials as well as concerns around additional cost ▪Potential for people to push poor quality PwRC or materials from a sustainability viewpoint					High standards may foster complacency, as stakeholders view compliance as sufficient without pursuing broader sustainability goals High standards often serve higher-end markets only, potentially excluding more accessible, widespread sustainability solutions PwRC lacking certification may not earn points, and rating tools may fail to incentivise PwRC use, reducing the overall policy impact	

Source: Authors.

Table 5. Stakeholder's responses to the negative impacts of policies with indirect impacts

Policies/ Impacts	Landfill tax	Extended producer responsibility	Penalties for illegal waste dumping	Restriction on the Disposal of Priority Waste	Accelerated Permission for On-site Recycling	Demolition Deposit/Refund	Proximity Principle	Ban on the Export of C&D Waste Resources	Product Stewardship	Levy waiver on waste recycling residuals
Impact 1	Non-uniform levies may lead to waste being transported to low-levy areas, avoiding proper waste management in high-levy regions	Many EPR schemes in construction are theoretical only, with limited practical application, as post-use recycling often isn't feasible due to contamination	Illegal dumping and disposal practices continue despite regulatory efforts, undermining waste management initiatives	Significant effort is needed to safely utilise potentially hazardous materials, with consideration of long-term environmental impacts	Use of contaminated materials on construction sites, due to lower costs, poses safety and environmental risks if not properly managed	Businesses with limited cash flow may resist implementing the policy due to the financial strain it imposes, potentially reducing overall compliance	The policy may not be feasible for regional areas where access to recycling facilities and resources is limited	Waste materials continue to be illegally dumped, bypassing proper disposal channels and causing environmental issues	Companies may engage in greenwashing, giving a misleading impression of environmental responsibility without significant action	Some recycling facilities may not efficiently recycle all materials they receive, reducing the overall impact of recycling policies
Impact 2	Higher landfill levies incentivise illegal dumping and disposal, increasing environmental risks	Risk of greenwashing, with schemes possibly altering reported volumes of PwRC to appear more effective	The policy inadvertently encourages the establishment of concealed dumping locations to avoid detection and regulatory fees	Large corporations may lobby regulators or governments for exemptions, which could undermine policy effectiveness	Not all projects have the capacity or resources to effectively implement this policy, reducing its overall impact		The policy might unintentionally incentivise illegal dumping as businesses seek to avoid costs and restrictions	Policies may incentivise avoiding transport to regions with better or cheaper treatment, leading to inequities in waste management	Poorly designed product stewardship can result in cost leakage, allowing free-riders and rogue operators to avoid financial responsibilities	It does not fully eliminate the production of residual waste, requiring continued disposal solutions
Impact 3	Incorrect categorisation, such as labelling contaminated materials as "Clean," can lead to cross-contamination and hidden hazardous materials (e.g., asbestos)	Inadequate EPR schemes can lead to cost shifting to non-responsible parties and create 'free-rider' problems with unregulated operators		The policy could drive an increase in illegal dumping activities as entities seek to bypass legal disposal methods	There is a risk that PwRC may not meet minimum quality requirements, potentially compromising construction integrity		Treating recyclable materials as waste can undermine recycling efforts and result in lost opportunities for reuse	Certain recyclable wastes may not be processed due to a lack of local facilities capable of handling them, limiting recycling potential	Stewardship organisations must monitor the quality of materials processed to ensure the PwRC meets necessary standards	Policies are open to abuse, allowing some operators to exploit recycling schemes without contributing to genuine environmental outcomes

Policies/ Impacts	Landfill tax	Extended producer responsibility	Penalties for illegal waste dumping	Restriction on the Disposal of Priority Waste	Accelerated Permission for On-site Recycling	Demolition Deposit/Refund	Proximity Principle	Ban on the Export of C&D Waste Resources	Product Stewardship	Levy waiver on waste recycling residuals
Impact 4	Landfill levies encourage stockpiling, particularly of C&D waste in regions like WA, where demand and supply chains are insufficiently developed	Challenges with accessing global spare parts and shortages of skilled labour may hinder effective implementation		Policies may inadvertently encourage the development of a black market for illegal dumping to avoid fees and regulations	Fast-tracked projects may overlook necessary safety considerations, which could result in long-term environmental and structural issues		Waste continues to be sent to Qld, highlighting enforcement challenges and difficulty in administering cross-border regulations	Insufficient capacity in waste management facilities results in improper handling of waste materials	Policies and stewardship schemes can increase costs for businesses, potentially impacting compliance and profitability	•Some businesses may falsely claim recycling activities to receive levy waivers, especially in construction and demolition where material weight makes it financially beneficial •Certain entities may engage in tax avoidance strategies to circumvent recycling levies and reduce costs
Impact 5	Local governments may see an increase in illegal dumping due to levy avoidance tactics	Placing full life-cycle responsibility on manufacturers is complex and may not be easily achievable		Strict policies may restrict innovation in recycling and reuse methods, limiting advancements in sustainable practices	Policies are not yet applied at a large enough scale to achieve optimal efficiency and effectiveness		The current economic setup may not fully support the policy's recycling and CE initiatives, limiting its effectiveness	Banning exports without restricting imports results in accumulating low quality materials domestically, worsening the waste problem	Shortened stewardship cycles may result in reduced material quality, as procures rush to meet policy timelines	
Impact 6	Higher levies might also encourage the illegal export of waste materials	Due to the long lifespan of construction materials (10-20 years), it can be easy to neglect responsibilities over time			Australia's status as a net importer limits the impact of local recycling efforts		The need for proximity to recycling facilities may limit options for projects, particularly those outside the policy's boundary, stifling some opportunities	Contaminants, such as asbestos, need to be managed carefully to prevent health hazards in PwRC	Low production capacity in Australia limits the effectiveness and scope of product stewardship schemes due to constrained supply	

Table 5. (Continued)

Policies/ Impacts	Landfill tax	Extended producer responsibility	Penalties for illegal waste dumping	Restriction on the Disposal of Priority Waste	Accelerated Permission for On-site Recycling	Demolition Deposit/Refund	Proximity Principle	Ban on the Export of C&D Waste Resources	Product Stewardship	Levy waiver on waste recycling residuals
Impact 7	Encouraging PwRC use without careful consideration may lead to the use of less beneficial or unsuitable PwRC	Policies may result in higher overall construction costs, impacting budgets and project viability			•Potential for contamination, especially if recycling activities occur onsite, leading to local environmental issues •Policies can lead to local impacts, such as increased noise, dust and potential safety hazards, affecting surrounding communities. •Local residents may experience inconvenience from policy impacts, including noise, odour and dust from nearby construction or recycling activities		The policy may be inappropriate for certain areas, such as highly developed suburbs, where recycling and waste management infrastructure is limited or unsuitable	Contaminated waste may be exported illegally due to high domestic treatment costs, circumventing regulations	Limited availability of PwRC restricts the ability to meet demand, impacting policy success	
Impact 8	The balance of costs (levy vs. incorporating recycled content) impacts compliance, with higher landfill taxes increasing the motivation for illegal disposal	In Australia, many EPR schemes are voluntary, apply only to select products, and thus have limited impact					•Bans on certain waste exports can lead to unintended consequences, such as increased stockpiling and local landfill use •The industry struggles to manage the volume of waste generated, leading to stockpiling due to insufficient processing infrastructure		Some policies actively hinder the reuse of materials, especially when dominated by new material producers who may design products to limit recyclability	
Impact 9	Levy structures inadvertently push certain lightweight materials, like textiles, towards landfilling due to a lack of recycling incentives	These policies can create a false perception that the waste problem is solved, while also shifting landfill and recovery costs from councils back to manufacturers					•Without local markets for PwRC, waste may end up in landfills or be illegally dumped, increasing environmental risks. •The underdeveloped Australian market struggles to absorb all C&D waste generated, leading to difficulties in reclaiming valuable materials. •Without sufficient market demand, recycling may not occur, as global warming solutions require international collaboration			

Policies/ Impacts	Landfill tax	Extended producer responsibility	Penalties for illegal waste dumping	Restriction on the Disposal of Priority Waste	Accelerated Permission for On-site Recycling	Demolition Deposit/Refund	Proximity Principle	Ban on the Export of C&D Waste Resources	Product Stewardship	Levy waiver on waste recycling residuals
Impact 10		The policy affects producers differently depending on their products, leading to an imbalance and potentially unfair competition								

Source: Authors.

Table 6. Key measures to enhance sustainable procurement effectiveness

No	Measure	Description	No	Measure	Description
1	PwRC application targets	• The policy should set clear priority material targets to improve local waste management and align industry efforts with policy intent • Specific local-level targets for PwRC adoption should be established to drive consistent progress	11	A wider industry engagement	Greater engagement with the construction sector beyond road projects should be prioritised in Western Australia.
2	Application of PwRC certification	Prioritise PwRC certified by recognised sustainability standards, such as the Forest Stewardship Council (FSC) or LEED. This ensures transparency and credibility in the sourcing process	12	Life cycle benefit analysis	Life Cycle Benefit Analysis (LCBA) should be conducted to compare the environmental impacts of PwRC and virgin materials.
3	Increase awareness and educational programs	• Industry awareness campaigns should promote the use of PwRC in construction projects • Educational programs should highlight the economic and environmental benefits of using PwRC • Awareness campaigns should inform construction companies and the public about PwRC benefits	13	Procurement requirements	PwRC content requirements should be included in tenders, and procurement rules should be standardised across states.

Table 6. (Continued)

No	Measure	Description	No	Measure	Description
4	Financial incentives	• Financial incentives such as tax credits should be provided to encourage PwRC adoption in construction • Financial incentives and subsidies should be provided to companies that aim to transition to supply and use PwRC	14	Collaboration and coordination	Stakeholder collaboration should be improved to enhance resource recovery and circular waste management.
5	Regulatory support	Supportive regulations should be implemented to promote the use of PwRC in the construction industry	15	Risk and contamination concerns	Contamination standards should be established to build confidence in PwRC materials.
6	Industry leadership	Organisations should adopt PwRC to lead by example and stay competitive within industry standards	16	Life cycle and client involvement	Clients should play a leadership role in waste reduction and promote the use of PwRC in building projects.
No	Measure	Description	No	Measure	Description
7	Local government role	Local governments should prioritise PwRC in procurement tenders and support local suppliers	17	Supplier engagement	PwRC material suppliers should be engaged to ensure a reliable and consistent supply chain.
8	Cost of externalities	Policies should account for external costs like carbon footprint when comparing virgin materials and PwRC	18	Monitoring and reporting	Mechanisms for monitoring and reporting PwRC usage and sustainability progress should be implemented.
9	Standardisation	Standards should be developed to make PwRC a default choice rather than an optional one	19	Addressing industry hesitance	Research and cost analysis should address industry hesitance and demonstrate the value of PwRC in construction.
10	Voluntary to mandatory policy	PwRC policies should transition from voluntary to mandatory to ensure industry-wide compliance.			

Source: Authors.

Table 7. Key measures to enhance recycled product certification effectiveness

No	Measure	Description	No	Measure	Description
1	Policy implementation	Policies should enhance transparency and establish standards for PwRC	15	Research and innovation support	Support R&D to improve PwRC quality and environmental performance
2	Market confidence	Efforts should boost market confidence in PwRC, ensuring better access and acceptance	16	Stakeholder collaboration	Strengthen partnerships between government, industry, and certification bodies
3	Certification in WA	Increase availability of certified products in Western Australia	17	Australian testing standards	Develop Australian testing standards for PwRC similar to conventional materials
4	Awareness of the benefits of RPC schemes	• Industry awareness campaigns should highlight the benefits of certification programs • Focus education on increasing the industry's understanding of PwRC certifications • Conduct campaigns to educate end-users about the benefits of certified PwRC	18	Imported PwRC concern	Prevent reliance on cheaper imported PwRC over local PwRC Ensure certifications don't incentivise cheaper imports over local recycled products
5	Case studies on PwRC	Develop case studies to showcase certified PwRC success in projects	19	Testing requirements	Align testing requirements for PwRC with Australian standards
6	Compliance requirements	Set thresholds for the use of certified PwRC, e.g., 25% GreenTag products in projects	20	Alignment with existing standards	Certification programmes should align with existing regulatory and green building standards.
7	Certification confidence	Certifications should provide consumer confidence and competitive differentiation	21	Avoiding negative impacts	Ensure PwRC certifications account for potential environmental risks
8	Global certification expansion	Expand programmes like GGT globally for better access to certified PwRC	22	Digital passports and traceability	Implement digital passports to enhance traceability of recycled content
9	Legitimacy of PwRC	Establish RPC as a credible and recognised standard	23	Broader certification criteria	Expand certification to include more building materials and establish clear performance standards.

Table 7. (Continued)

No	Measure	Description	No	Measure	Description
10	Certification incentives	• Provide financial incentives to encourage certification adoption • Introduce financial incentives like tax reductions to promote certified PwRC adoption • Provide government funding or low-interest loans to help businesses obtain certifications • Provide subsidies to help SMEs achieve certifications	24	Standardised certification	• Certification criteria should be standardised across suppliers and policymaking entities. • Certification criteria should be harmonised across regions for ease of adoption. • Align different certification schemes (e.g., GGT, GECA and C2C) for easier market adoption.
11	Integration with green standards	Integrate RPC into existing green standards like LEED	25	Policy requirement for certification	• Policies should mandate certification and traceability for PwRC to improve reliability. • Mandate certifications to ensure PwRC market adoption
12	Procurement specifications	• Public/private procurement policies should mandate certified PwRC • Mandate certification for public projects to ensure quality and sustainability	26	Government/scientific endorsement	Certifications should be backed by government or scientific bodies for
13	Capacity building	Train manufacturers on certification processes	27	RPC scheme coverage issues	Address inconsistencies in RPC schemes implementation across Australia
14	Transparency and traceability	• Develop mechanisms to ensure certified PwRC legitimacy in the supply chain • Enforce chain-of-custody to verify recycled content claims	28	Mitigating greenwashing	• Use certifications to reduce greenwashing risks and increase product transparency. • Use third-party certifications to enhance product credibility and reduce greenwashing.

Source: Authors.

Table 8. Key measures to enhance tax on virgin materials effectiveness

No	Measure	Description	No	Measure	Description
1	Managing cost implication	• Implementation should avoid adding excessive costs to the economy • Balance the cost between PwRC and non-PwRC supply chains to encourage adoption • Adjust tax rates to create financial incentives for PwRC over virgin materials • Apply Virgin Tax to reduce reliance on virgin plastic and boost PwRC demand	14	Increase Australian product use	Use Virgin Tax to favour Australian PwRC over cheaper imports Avoid making Australian virgin materials more expensive, which could encourage imports with higher carbon footprints.
2	Price signal for RCA	Establish price signals favouring Recycled Concrete Aggregate (RCA) over quarried aggregates	15	Market-based instruments	Use tools like tradable permits or carbon pricing to internalise the environmental costs of virgin materials.
3	Awareness of reporting requirements	Increase awareness of reporting, weight calculation, and claiming relief	16	Education and awareness	Launch campaigns to educate industry on PwRC benefits and Virgin Tax implications.
4	Virgin materials tax application	Impose heavy taxes on natural materials to reduce PwRC costs and counter perceptions of high PwRC prices Apply Virgin Tax to specific materials to drive PwRC supply chains	17	Industry collaboration	Collaborate with industry to set standards and best practices for PwRC use.
5	Tax definition clarity	Use "raw virgin materials tax" to distinguish from excavated materials, encouraging virgin excavated natural material (VENM) reuse	18	Certification and verification	Implement certification to ensure PwRC quality and build market trust.
6	Environmental protection	Use taxation to prevent environmental destruction Ensure taxes consider environmental footprints, regulatory environments, and transport impacts	19	Research and development	Fund R&D for recycling innovations, improving PwRC performance and cost-effectiveness.

Table 8. (Continued)

No	Measure	Description	No	Measure	Description
7	International agreements	Verify international material supply chains for consistency and accountability	20	Standards for PwRC	Develop standards to define PwRC applications in construction.
8	Graduated tax rates	Set graduated tax rates based on recycled content; taxing virgin materials higher	21	Public procurement policies	Require public projects to prioritise PwRC and demonstrate compliance.
9	Tax revenue allocation	• Allocate Virgin Tax revenue to support the recycling industry and PwRC development • Allocate tax revenue to promote PwRC development and market visibility • Use Virgin Tax revenue to fund R&D for new technologies and applications • Ensure Virgin Tax revenue supports circular economy processes and nature-based solutions • Use tax revenue to cover transport costs and support waste processors	22	Tax incentives for PwRC use	• Provide tax incentives or exemptions for projects using PwRC • Offset costs to incentivise PwRC adoption and support businesses investing in sustainable materials • Offer tax credits or rebates to incentivise PwRC investment
10	Shift from virgin sources	Shift reliance from virgin materials to PwRC as the default option	23	Budget constraints and greenwashing	Address low-budget project challenges and minimise risks of greenwashing with minimal recycled content.
11	Controlled quality supply	Maintain quality control to ensure PwRC can effectively replace virgin materials	24	Performance and longevity	Consider cases where virgin materials offer better durability and sustainability before taxing.
12	Regulation and black-market risks	Address potential black-market risks and regulatory feasibility	25	Recycled asphalt and cementitious materials	Encourage recycled asphalt and cementitious materials through Virgin Tax.
13	Underpriced virgin materials	Highlight how under-pricing virgin materials distorts market comparisons, making PwRC seem expensive	26	Tax rate sufficiency	Ensure Virgin Tax rates are high enough to incentivise PwRC use over virgin materials.

Source: Authors.

Table 9. Key measures to enhance financial incentive policy effectiveness

No	Measure	Description	No	Measure	Description
1	Traceability system	Implement a traceability system to ensure transparency and reliability in PwRC use	19	Industry standards and collaboration	Industry collaboration on PwRC standards ensures quality and consistency.
2	Price incentive clarification	Clarify specific price incentive policies; ensure they address key sectors like road infrastructure	20	Guidelines for PwRC use	Develop construction guidelines for safe and effective PwRC applications.
3	Encouraging new users	Design policies to encourage first-time users to try PwRC Provide funding for initial infrastructure costs related to PwRC adoption	21	Crumb rubber on roads	Support trials and adoption of higher crumb rubber content in road projects.
4	Targeted financial incentives	Provide tailored financial incentives for sectors managing municipal, C&D, and general waste	22	Prioritisation for high-quality PwRC	Ensure incentives are tied to high-quality PwRC to build confidence.
5	Market efficiency and new markets	Improve market efficiency and foster new markets for PwRC through policy frameworks	23	Life-cycle incentives	Apply financial incentives throughout the project lifecycle to increase PwRC usage.
6	Central database	Develop a centralised database for PwRC procurement to improve transparency and accessibility	24	Awareness through value proposition	Position PwRC as a cost-saving and socially beneficial value-add to drive adoption.
7	Client involvement in sustainability	Increase client involvement and leadership in promoting PwRC and sustainable construction Project clients increasingly demand recycled content in materials	25	Reduced processing costs	Lower processing and logistics costs to further encourage PwRC use Provide monetary incentives to mitigate high processing costs and stimulate PwRC supply chains
8	National adoption	Extend PwRC policies and incentives nationwide to standardise impact	26	Contamination control	Verify contamination levels in PwRC to enhance confidence in incentivised materials.
9	Targeting underutilised materials	Focus on underutilised materials to bolster supply chain sustainability	27	Long-term viability	Ensure PwRC remains cost-effective in production for sustainable long-term adoption.
10	Incentives for green materials	Expand financial incentives to include all green materials and methods	28	Local content priority	Prioritise locally produced recycled content to support local supply chains.

Table 9. (Continued)

No	Measure	Description	No	Measure	Description
11	Increased financial benefits	Develop incentive schemes that provide meaningful value to stakeholders Boost subsidies, tax breaks, and grants for PwRC projects	29	Complement virgin tax	Pair financial incentives with virgin material bans for effective circularity. Combine incentives with a Virgin Tax to drive sector-wide change.
12	Mandatory recycled content	Mandate minimum recycled content in construction materials to drive PwRC usage	30	Risk of subsidy dependency	Address risks of subsidy dependency and potential budget issues post-withdrawal.
13	Public awareness campaigns	Launch campaigns to educate stakeholders about the benefits of PwRC	31	Differentiated pricing approach	Implement pricing mechanisms to prevent favouring heavy materials over lighter ones like insulation.
14	Alternative incentives	Explore alternative incentives beyond monetary rewards to ensure long-term sustainability	32	Encouraging private sector engagement	Extend incentives to encourage private sector participation in PwRC adoption.
15	Streamlined certification	Simplify PwRC certification to lower adoption barriers	33	Budget constraints and greenwashing	Manage low-budget project impacts and greenwashing risks by ensuring meaningful recycled content.
16	Supplier collaboration	Collaborate with PwRC suppliers for a reliable and consistent supply chain	34	Independent scientific studies	Support peer-reviewed studies on PwRC to build public trust and legitimacy and avoid conflicts of interest.
17	R&D funding	Investing in recycling innovation and new PwRC technologies broadens the range of materials available for construction	35	Case studies and benchmarking	Sharing case studies and promoting benchmarking of best practices can support PwRC uptake.
18	Government procurement policies	Mandate PwRC prioritisation in government projects to increase demand			

Source: Author.

Table 10. Key measures to enhance the effectiveness of harmonised PwRC specifications

No	Measure	Description	No	Measure	Description
1	Cross-jurisdictional use	Policy relevance increases when PwRC can be used across multiple jurisdictions	16	Communication with stakeholders	Standards must be clearly communicated to ensure stakeholder understanding and compliance.
2	Performance-based standards	Performance-based standards should allow flexibility in PwRC use while meeting functional requirements	17	Support for suppliers	Provide funding support for suppliers and manufacturers to overcome adoption barriers.
3	User confidence through awareness	Raising awareness of standards boosts user confidence in PwRC	18	Avoiding the lowest standard default	Avoid standardisation defaults that reduce existing PwRC use
4	Circular economy integration	Embed circular economy principles to encourage design for disassembly, reuse, and recyclability	19	R&D investment	Fund R&D to improve the quality, performance, and cost-effectiveness of PwRC
5	Minimum standards	Set minimum standards for PwRC to improve industry quality outcomes	20	Capacity building and training	Provide training for construction professionals on PwRC use and compliance
6	Life cycle assessment	Integrate LCA to ensure PwRC provides environmental benefits	21	Harmonised legislation	Harmonise waste legislation and introduce smart waste models
7	Showcasing for adoption	Early adoption can showcase policy success and encourage broader adoption across Australia	22	Monitoring and evaluation	Establish monitoring and evaluation to refine PwRC specifications
8	Standardisation and certification	• Standardise testing and certification to build confidence among construction professionals • Promote certification and benchmarking to ensure PwRC quality and compliance	23	Expansion to other materials	• Harmonised specifications should cover a broad range of materials without stifling innovation. • Develop specifications for a wider range of construction materials that can use PwRC, e.g., concrete, steel, and glass.
9	PwRC performance standards	Demonstrate that PwRC performance matches or exceeds non-PwRC materials through demonstration projects	24	Focus on ease of business	Consistent standards simplify business and facilitate PwRC adoption
10	Market development	Stimulate the PwRC market through procurement policies, incentives, and awareness campaigns	25	Focus on ease of business	Consistent specifications should simplify business and facilitate PwRC adoption.

Table 9. (Continued)

No	Measure	Description	No	Measure	Description
11	Government leadership and action	• Early government adoption can lead to broader industry uptake of PwRC • WA's Roads to Reuse shows the benefits of high-quality, contamination-free PwRC	26	National consistent information-sharing	National standards and consistent information-sharing are essential for effective implementation.
12	PwRC standards development	Developing standards that reduce risks and encourage end users to apply PwRC	27	Legislative approval for mixed content	Legislate for mixed recycled materials in traditional materials to boost adoption.
13	Encouraging innovation	Specifications should foster innovation and material development	28	Alignment with existing standards	Align policies with current certification and manufacturing standards
14	Cross-jurisdictional cooperation	Industry collaboration is needed to harmonise specifications across states	29	Supply chain consistency	Ensure strong and consistent supply chains to support PwRC adoption across regions.
15	Expert confidence	Specifications should increase confidence among engineering professionals, encouraging PwRC adoption.			

Source: Authors.

Table 11. Key measures to enhance the effectiveness of environmental sustainability rating

No	Measure	Description	No	Measure	Description
1	Rating scheme encouragement	Use rating schemes and tools to drive wider PwRC use, especially with specific material targets	16	Monitoring and feedback mechanisms	Establish feedback systems to continuously refine PwRC promotion strategies in the construction.
2	Demand for standards	Create demand for IS, Green Star, and LEED standards to motivate organisations to meet benchmarks	17	Focus on Australian PwRC	Emphasise Australian PwRC in rating tools, excluding imported materials from credits

No	Measure	Description	No	Measure	Description
3	Other rating systems	Develop alternative rating systems to further environmental sustainability	18	Public awareness and case studies	Promote case studies showcasing high-standard projects to raise industry awareness to aim for higher.
4	Whole-of-life approach	• Increase focus on life-cycle impacts, including building operations to enhance sustainability outcomes • Adopt life-cycle thinking that supports PwRC but also encourages reuse or refurbishment over new production	19	Addressing multiple rating tools' complexity	• Simplify multiple rating tools to avoid consumer confusion and improve effectiveness. • Simplify administrative requirements to encourage companies to exceed minimum standards.
5	PwRC-specific criteria within ratings	• Add specific PwRC criteria to rating systems, rewarding PwRC use with extra points or credits • Increase PwRC integration within the IS tool and expand its relevance in Green Star ratings • Encourage specific PwRC criteria in rating systems to incentivise the use • Address PwRC's minor role in current environmental rating schemes to drive its expansion	20	Managing the risk of added costs	• Mitigate risks of additional PwRC sustainability rating costs that could increase project expenses. • Provide subsidies to make rating certifications accessible to smaller organisations. • Develop effective rating options that enable all developers to follow sustainability goals.
6	Increasing ratings requirements and targets	• Spur product innovation by increasing sustainability ratings and targets • Set objectives beyond minimum rating targets to encourage PwRC adoption	21	Legislation for significant projects & standard practice	• Enforce sustainability rating compliance in significant projects or high-value contracts through legislation • Contracts requiring Specific rating level requirements that make the use of PwRC a standard practice
7	Educational outreach for PwRC	Provide resources to guide professionals in sourcing PwRC aligned with rating systems	22	Innovation and research promotion	Reward innovation in PwRC materials and technologies through rating system incentives.

Table 11. (Continued)

No	Measure	Description	No	Measure	Description
8	Market access for alternative products	Expand market access for alternative products, offering sustainable more construction options	23	Limited products for rating schemes	Address limited availability and high costs of rated products, which hinder smaller manufacturers.
9	LCA expansion	Incorporate LCA principles into rating systems to evaluate environmental impacts from extraction to disposal	24	Addressing point-chasing risks	Prevent point-chasing in rating systems to ensure substantive sustainable practices.
10	Prioritising high-recycled content materials	Encourage builders to prioritise high-recycled-content materials in projects	25	Rating schemes' cost as a quality filter	High certification and sustainability rating costs often reflect a genuine commitment to sustainability.
11	Transparency and reporting	Require transparency and reporting on PwRC use as part of the rating process to promote accountability	26	Project-specific implementation	Align rating scheme use with organisational and policy goals for consistent impact.
12	Generating market demand	Use ratings to create market demand for PwRC by highlighting sustainability benefits	27	Voluntary and process-driven ratings	Address limitations of voluntary process-driven ratings- which may limit circular or regenerative benefits- by focusing on outcomes.
13	Supplier engagement	Motivate suppliers to certify PwRC and validate its environmental attributes	28	Corporate social responsibility recognition	Leverage Corporate Social Responsibility (CSR) recognition in ratings to boost PwRC adoption across the built environment
14	Rewarding sustainable projects	• Reward projects incorporating new PwRC technologies, fostering recycling advancements • Offer financial incentives for projects achieving high sustainability ratings through PwRC use	29	Policy alignment	• Align rating systems with government policies to support PwRC adoption • Align policies with Green Star, IS, and LEED to ensure rating system adoption without relying solely on voluntary adoption
15	Expanded scope for building ratings	Expand building rating scope to cover more types, increasing PwRC use			

Source: Authors.

Table 12. Key measures to enhance the effectiveness of carbon tax policy

No	Measure	Description	No	Measure	Description
1	Supplier availability	Ensure the availability of suppliers offering PwRC with lower embodied carbon	16	Target application areas for low-carbon materials	Focus policies on areas where low-carbon materials are viable, excluding critical applications where PwRC are less suitable
2	Carbon Tax Scheme awareness	• Increase understanding of carbon tax or credit schemes to incentivise PwRC use by reflecting emissions costs • Educates stakeholders about the carbon benefits of PwRC and Carbon Tax Schemes to build support	17	Co-investment in innovation and recovery infrastructure	• Advocate for government co-investment to foster innovation in recycling processes and material development • Invest in recycling infrastructure to improve PwRC supply
3	Tax incentives and financial support for PwRC use	• Monetise PwRC to encourage its adoption in the building and construction sector • Offer tax rebates or incentives for companies using PwRC to offset higher sourcing costs • Provide financial assistance to offset initial PwRC adoption costs in construction	18	Lifecycle carbon pricing	• Cover all lifecycle emissions in carbon pricing to ensure comprehensive environmental impact accounting and recycling benefits are not limited • Advocates for lifecycle impact assessments rather than carbon-only metrics to avoid unintended consequences • Full lifecycle carbon pricing is most effective, accounting for impacts from vegetation loss for virgin material extraction • Recognise that PwRC manufacturing may have higher carbon due to additional processes, making lifecycle assessments critical
4	Commercial advantage	• Highlight competitive advantages for companies specifying and procuring PwRC • Emphasise competitive advantage and futureproofing as key motivators, beyond financial gain	19	Carbon tax reception	Address negative perceptions of carbon tax by providing viable alternatives, a 'tax' can cause negative perceptions.

Table 12. (Continued)

No	Measure	Description	No	Measure	Description
5	Tool limitations	Address concerns about carbon credits by developing quantitative tools	20	Lifecycle carbon pricing	Implement full lifecycle carbon pricing, including impacts from vegetation loss for virgin material extraction
6	Market creation through awareness	Raise awareness of PwRC and best practices to help establish a market	21	Local governments' Carbon Tax Schemes	Encourage local governments to apply their carbon costs, promoting sustainable choices as seen in Dutch Local Government Areas (LGAs)
7	Minimal difference for aggregates	Highlight that limited carbon differences between recycled and virgin aggregates increase PwRC appeal	22	Addressing the risk of using substandard materials	Mitigate risks of promoting low-quality recycled options by ensuring accurate lifecycle carbon calculations
8	Differentiated carbon pricing	Introduce carbon pricing based on emissions from virgin vs PwRC to incentivise PwRC use	23	External cost integration	Integrate external costs of virgin materials into pricing to reflect true costs and encourage PwRC use
9	Carbon credit incentives	Offer carbon credits for PwRC use, rewarding lower-carbon projects	24	Certification and verification	Certification systems should be utilised to ensure the quality and authenticity of PwRC
10	Transparent carbon footprint reporting	Provide transparent reporting on carbon footprints for construction materials to empower informed choices	25	End carbon offsetting	Advocate for ending carbon offsets in favour of direct legislation promoting PwRC
11	R&D funding for PwRC	Allocate R&D funding to reduce PwRC's carbon footprint, increasing competitiveness against virgin materials	26	EU and global carbon pricing models	Learn from global carbon pricing models to ensure effective implementation in Australia and avoid unwanted consequences
12	Market-based carbon instruments	Considers market-based tools (e.g., carbon trading) to reward emissions reductions from using PwRC	27	The price of carbon determinant	Set an effective carbon pricing rate to drive meaningful changes; Australia's current rate is ~$75/tonne
13	Public procurement for PwRC	Use public procurement policies to prioritise PwRC and set an example for the private sector	28	Carbon pricing and clean technology	Pair carbon pricing with mandatory climate reporting to drive innovation and clean technology development

No	Measure	Description	No	Measure	Description
14	Carbon accounting standardisation	Standardise carbon accounting for recycled and virgin materials through industry collaboration Address complexities in assigning embodied carbon responsibility (developers vs asset owners)	29	Carbon tax revenue redistribution	Hypothecate carbon tax revenue to promote market development and sustainability Carbon tax reinvested in innovation and recycling
15	Long-term policy stability	Ensure policy consistency to encourage sustainable investment and reliable PwRC supply chains			

Source: Authors.

Table 13. Key measures to enhance the effectiveness of landfill tax policy

No	Measure	Description	No	Measure	Description
1	Application scope	• Inconsistencies in levy application undermine its effectiveness; uniform application is crucial • Inconsistent levies across states reduce overall effectiveness compared to uniform application	13	Sustainable procurement	Public sustainable procurement prioritising PwRC increases demand and sets an example for private-sector adoption.
2	Local supply chains support	Develop local recycling supply chains to facilitate sustainable PwRC use and reduce landfill demand	14	Address levy avoidance and stockpiling	Address levy avoidance and stockpiling by pairing this policy with sustainable procurement can build PwRC markets
3	Levy rate	• Higher levies can lead to illegal dumping, requiring monitoring and regulation • Higher levies should be paired with increased recycling options to reduce incentives for illegal dumping • Increased costs for C&D waste disposal can create more recycling opportunities • Avoid excessive levies, taxes, and penalties, suggesting a preference for incentives	15	Education and awareness	• Clearly outline policy purpose to facilitate responsible industry decision-making. • Highlight that landfill levies are globally proven tools for market development for recycling and circularity. • Engage construction stakeholders to promote PwRC benefits through knowledge-sharing, training, and showcase projects.

Table 13. (Continued)

No	Measure	Description	No	Measure	Description
4	Investment in recycling infrastructure	• Invest in recycling infrastructure to support PwRC markets • Raising levies without increasing recycling options can lead to continued landfill use at a higher cost • Providing subsidies for land and operations can stabilise the recycling industry and supply chains	16	Paired with circular economy investment	• Pair levies with circular economy opportunities reduce risks of illegal dumping and support sustainable practices. • Levy funds should support recycling infrastructure and innovation to drive circular economy development.
5	Waste diversion and job creation	Diverting waste from landfills creates sub-industries, job opportunities in upcycling, and supply chain development	17	Target downstream supply chain	Landfill levies influence downstream supply, potentially increasing PwRC supply.
6	Certification and recognition	Establish a certification system that assesses the lifecycle impact of PwRC materials to help builders prioritise environmentally friendly options	18	Regular levy review	Levy should be regularly reviewed to ensure it effectively discourages landfill disposal.
7	Levy revenue redistribution	• Allocate levy revenue to recycling infrastructure, supporting PwRC processing • Levy revenue should support recycling facility expansion and PwRC market development • Levy rebates should incentivise waste reduction and recycling	19	Policy linkage to holistic waste framework	Ensure that landfill levies are part of broader waste and recycling policy frameworks to prevent unintended behaviours.
8	Market development	Increase market demand to address large quantities of unused recycled aggregate	20	Surcharge for recyclable content in landfill deliveries	Apply surcharge on landfill deliveries containing recyclables can incentivise pre-sorting and recycling efforts
9	Levy and circular economy	• Landfill levies are critical for driving recycling and circular economy adoption • Landfill levies are a critical driver of recycling and circular economy adoption, with a significant impact on behavioural change	21	Avoid passing levies to consumers	• Levies should be borne by waste producers, not consumers, to drive industry change. • Ensure levy costs are not transferred to consumers, promoting EPR to hold waste producers accountable.

No	Measure	Description	No	Measure	Description
10	Incentives over taxes	• Incentives are more effective than landfill taxes in encouraging recycling • Avoid excessive levies and taxes; incentives are preferred • Provides tax rebates to incentivise PwRC use, reducing raw material demand and landfill	22	Lightweight material challenges	Consider material weight in levy structure; lightweight materials are often easier to landfill, reducing recycling incentives for some industries.
11	Support R&D for PwRC	Offer grants for PwRC R&D to improve quality and durability, fostering innovation in PwRC	23	Enforcement of compliance	Effective compliance monitoring is essential to prevent levy avoidance and ensure levy payments support recycling objectives.
12	Mandatory PwRC targets	Set minimum PwRC usage standards to create demand and drive recycling innovation.			

Source: Authors.

Table 14. Key measures to enhance the effectiveness of EPR policy

No	Measure	Description	No	Measure	Description
1	Stakeholder inclusion	• Ensure all stakeholders share responsibilities based on contributions, avoiding imbalanced costs • Responsibility should be distributed among all lifecycle stakeholders, not just manufacturers • Recognise suppliers' significant influence on material selection sharing responsibility	14	Policy applicability in construction	• Acknowledge EPR's conceptual effectiveness but note its limited practical application and low recovery rates. • Highlight EPR challenges in construction due to long product lifecycles but note its value in other industries. • Address challenges in tracking responsibilities over long lifecycles (10-50 years) in the building and construction sector.
2	Supply-side focus	Further policies are needed to influence demand rather than supply	15	Mandated EPRs for the building sector	Mandate EPR in construction to support segregated waste streams and recycling.
3	Consumer awareness of EPR	Increase consumer awareness about producers' roles in product lifecycle, recycling, and disposal	16	Recycled content targets	Set specific PwRC targets within EPR requirements to encourage usage

Table 14. (Continued)

No	Measure	Description	No	Measure	Description
4	Case studies and best practices	• Share case studies and best practices to highlight successful product stewardship and EPR • Learn and adopt best practices from Europe in product stewardship and EPR	17	Enforce supplier compliance	• Introduce penalties to ensure procurers take policies seriously • Force suppliers and manufacturers to take responsibility for end-of-life impacts of their products
5	Whole lifecycle design	• Identify opportunities for sustainable materials and product reuse through whole-of-life design • Emphasises product design for reuse/recycling, including reverse logistics • Encourage manufacturers to implement take-back systems and design products for recycling	18	Focus on longer product lifespans	• EPR policies should encourage manufacturers to create materials with extended lifespans and multiple-use cycles • Incentivise durable, reusable, and improved supply chains by requiring suppliers to take back products.
6	Cultural shift needed	Counter the "out of sight, out of mind" mentality prevalent in affluent countries	19	Take-back and design for recycling	Encourage take-back systems and designs for recycling
7	Extending EPR scope	Extend EPR scope beyond packaging to cover other products	20	Demand creation for PwRC through product stewardship	Increase demand for PwRC by including them in product stewardship targets
8	Business pressure	Acknowledge that EPR may put pressure on businesses, posing potential implementation challenges	21	Segmentation of EPR in construction	Segment EPR in construction to address unique handling of different building materials at end-of-life
9	Accessibility and cost of logistics	Address logistics and costs that are key barriers to effective EPR implementation	22	Importer and free-rider prevention	Mandatory EPR should include measures to prevent freeloading by imports and ensure all suppliers share responsibility.

No	Measure	Description	No	Measure	Description
10	Scheme governance	• Ensure EPR schemes are well-governed and support real recycling costs while actively creating markets for PwRC • Governance improvements, suggesting academic, community, and governmental oversight for transparency and accountability	23	Robust tracking platforms	Implement rigorous tracking platforms for effective EPR scheme management
11	Container deposit scheme model	Implement a CDS-style refund model for a broader range of products	24	Encouraging CE principles	Embed circular economy principles throughout the product lifecycle
12	Limited impact on material selection	Address how future recyclability focus may not influence initial material choices	25	Lifecycle impact responsibility	Encourage companies to assess lifecycle impacts and take end-of-life responsibility.
13	Reporting and auditing	Require regular reporting and external audits to maintain compliance and trust in EPR systems.			

Source: Authors.

Table 15. Key measures to enhance the effectiveness of penalties for illegal waste dumping

No	Measure	Description	No	Measure	Description
1	Awareness raising	• Raise awareness among all stakeholders to prevent improper waste disposal • Launch public awareness campaigns to educate the construction sector on illegal dumping consequences and promote best practices • Fund campaigns educating stakeholders about PwRC benefits and proper waste disposal practices	10	Enforcement and penalty increase	• Increase penalties for illegal waste disposal to encourage responsible management and divert waste to recycling. • Strengthen enforcement through increased monitoring, inspections, and punitive measures to curb illegal dumping. • Strengthen penalties, including higher fines and licence suspensions to deter repeat offenders. • Use substantial penalties to deter offenders, avoiding the risk of penalties becoming just a business cost.

Table 15. (Continued)

No	Measure	Description	No	Measure	Description
2	Responsibility	Hold stakeholders accountable for waste disposal to ensure responsible practices	11	Monitoring and inspection	Implement systematic waste monitoring from collection to the final destination, with regular inspections for effective implementation.
3	EPA and licensing	• Advocate for EPA monitoring and licensing to strengthen regulatory frameworks • Significant regulatory enforcement is needed to address persistent illegal dumping issues	12	Improved enforcement in non-levy and regional areas	• Strengthen enforcement in regional and non-levy areas with improved waste transport and infrastructure • Address illegal dumping issues in regions, focusing on small companies and tradespeople
4	Reward system for reporting	Introduce a reward system to encourage reporting of illegal dumping incidents	13	Recycling and safety	Emphasise that a balance between recycling and safety is essential to prevent illegal dumping and promote recycling.
5	Industry collaboration	Collaborate with industry associations to promote compliance and responsible waste management	14	Illegal dumping linked to levy increases	Illegal dumping rises with increased waste levies when recycling options are limited, highlighting the need for broader support.
6	Technology solutions for surveillance	Invest in technology like surveillance cameras and drones can improve monitoring and catch offenders in real-time	15	Supply chain support for recycling	Develop recycling supply chains to encourage waste generators to choose recycling over landfilling.
7	Stricter penalties and sanctions	Strengthen penalties, including higher fines and licence suspensions, to deter repeat offenders	16	"Name and shame" as deterrent	Publicly identify offenders to deter illegal dumping and promote accountability.
8	Incentives for proper waste disposal	Offer tax breaks to construction companies practising responsible waste disposal	17	Support for LGAs	• Support LGAs with collection efforts, especially for hazardous or difficult-to-manage waste. • Provide adequate funding to local governments for better enforcement and waste management policy regulation.
9	Public-private partnerships	Foster collaborations between government, law enforcement, and private stakeholders to combat illegal dumping	18	Balance penalties with PwRC encouragement	Find an appropriate balance between penalties and PwRC incentives to generate improved compliance and PwRC use.

Source: Authors.

Table 16. Key measures to enhance the effectiveness of restrictions on the disposal of priority waste resources

No	Measure	Description	No	Measure	Description
1	Landfill levy	• C&D waste materials are rarely banned but levies and incentives could help push them away from landfill • While inert construction waste isn't usually banned from landfills, big levies can push it toward recycling facilities	10	Accelerated permission for recycling	Government support in approvals and fast-tracking licenses can help the recycling industry.
2	Financial incentives for recycling	• Offer tax credits and grants to encourage companies to adopt PwRC and achieve higher recycling rates • Use incentives to reduce landfill disposal, promote PwRC use and avoid illegal practices • Balance landfill bans with incentives to encourage sustainable practices	11	Flexible policies for innovation	Flexibility is necessary to allow innovative solutions that align with policy intentions, ensuring practicality and encouraging compliance.
3	Effective project design process	• Implement effective project design process and material selection • Promote modular and easily deconstructable materials to maximise recovery in construction and demolition projects.	12	Market development initiatives	• Invest in market promotion for PwRC, support demonstration projects, and run campaigns on PwRC benefits. • Supports the creation of local markets and dependable supply chains, essential for recycling hazardous and priority waste materials effectively
4	Higher value PwRC potential	Promote the use of higher-value PwRC to benefit both industry and the environment	13	Development of recycling facilities	Recycling infrastructure is necessary for effective sorting and separation before processing, diverting them from landfills effectively.

Table 16. (Continued)

No	Measure	Description	No	Measure	Description
5	Industry education and awareness	• Promote industry awareness and education on proper material handling • Increase industry awareness of their responsibility in waste management • Highlight that diverting waste from landfills supports environmental sustainability and generates economic benefits through PwRC sale	14	Hazardous waste management	• Encourage the industry to properly manage hazardous materials, diverting them from landfills. • Due to limited locations, valuable or hazardous waste materials often end up in general waste, necessitating broader awareness of proper disposal practices. • Use science-based restrictions to prevent hazardous waste from entering landfills and improve compliance. • Encourage organisations to embed disposal practices for valuable or hazardous materials within core operations.
6	National EPR implementation	• Federal intervention can enforce manufacturer responsibility, especially for toxic or long-lived chemicals in waste that local/state policies cannot effectively manage • Make manufacturers responsible for end-of-life disposal which may also incentivise PwRC adoption	15	Effective waste regulations	• Develop clear regulations that discourage free-riding and ensure competitors in the recycling sector follow responsible practices. • Differences in regulations for floor renovations versus new builds create loopholes, often leading to incomplete hazardous material assessments.
7	Potential for stockpiling	Address risks of stockpiling or illegal disposal when landfill bans lack complementary demand-side policies	16	Mandatory sorting and separation	Require on-site sorting and separation of construction waste to support recycling and reuse efforts.
8	Mandated use of PwRC in public projects	• Mandate PwRC use in public projects to boost demand and strengthen recycling infrastructure • Prioritise PwRC in government-funded construction projects to drive demand and support sustainable practices • Bans need to be enforced alongside policies mandating PwRC use to avoid unintended consequences	17	Incorporation of industry insight	• This policy should incorporate industry insights rather than solely serving political agendas to address underlying issues. • Industry collaboration with the government to establish high standards and consistent regulations ensures effective implementation of waste disposal policies.

Source: Authors.

Table 17. Key measures to enhance the effectiveness of accelerated permission for on-site recycling

No	Measure	Description	No	Measure	Description
1	Acknowledging limitations	Off-site processing may remain necessary in areas lacking sufficient space for on-site recycling	7	Innovation encouragement	Encourage innovative recycling concepts specific to construction sites to enable new solutions and supply chain expansion.
2	Circular design practice	Encourage recyclable materials in project designs and builds	8	Government-industry collaboration	Foster collaboration between industry and government to establish best practices for on-site recycling and sustainable waste handling
3	Quality assurance and certification	• Ensure qualified inspectors certify material quality to meet design specifications • Implement a system to ensure on-site generated PwRC meet quality standards would build confidence in their reuse • Perform quality assurance for PwRC to ensure their safety and compliance with design specifications, promoting trust in the construction industry's use of PwRC	9	Technical and financial support	• Provide financial incentives and technical guidance to encourage investment in recycling infrastructure. • Standardised guidelines for on-site recycling, supplemented with quality audits, can increase safe recycling operations. • Government support fosters sustainable practices and innovation in the industry. • Incentives for on-site recycling reduce operational costs, benefiting both businesses and the environment.
4	Investment in waste recovery infrastructure	• Boost confidence in recycling infrastructure investment to support market demand for PwRC • Provide low-interest loans or subsidies to promote recycling facility investment	10	Encourage local recycling to reduce emissions	Enable more local recycling options to cut emissions from material transport.
5	Process simplification	• Simplify processes to promote recycling, lower waste volumes, and improve waste management • Streamline processes to increase material availability and improve waste processing efficiency	11	Prioritise trusted organisations	• Fast-track approvals for trusted organisations that meet quality standards to lower entry barriers and increase recycling output • Fast-track permits for trusted companies and set guidelines for on-site checks to support PwRC growth. • Streamline approvals for trusted companies to retrofit existing sites to encourage broader adoption of recycling practices.

Table 17. (Continued)

No	Measure	Description	No	Measure	Description
6	Health and safety concerns	• Balance regulatory approaches to maintain safety while supporting efficient recycling operations • Understand and address work health and safety impacts to prevent unintended consequences • Subject on-site processes to EPA risk analysis to ensure compliance and minimise environmental impacts • Address noise and dust control to reduce community opposition to recycling facilities • Certification requirements for recyclers to ensure processes are safe and environmental impacts are minimised	12	Education and case studies	• Highlight successful examples to encourage adoption and sustainable practices across the industry. • Educate policymakers on how quicker approvals lead to lower set-up costs, encouraging companies to start recycling sooner. • Highlight that on-site recycling reduces costs and emissions by eliminating transport and double handling. • Communicate that streamlined processes can reduce costs and emissions, delivering economic and environmental benefits. • Demonstrate that faster approval processes for on-site recycling facilities, with incentives, could encourage the use of PwRC in construction.

Source: Authors.

Table 18. Key measures to enhance the effectiveness of demolition deposit-refund policy

No	Measure	Description	No	Measure	Description
1	Recycling and disposal awareness	• Raising awareness about recycling and disposal of C&D waste encourages early adoption of PwRC materials in project designs • Raise awareness of recycling locations, such as Community Recycling Centres, to prevent improper material disposal • Increasing awareness about material disposal can create more responsibility around waste management and reuse • Highlight that the use of incentives like deposit/refund schemes to support circular practices in construction	10	Combined application of CE policies	• Combine policies like demolition deposit schemes with other frameworks to strengthen waste management practices. • Integrate construction recycling incentives with circular economy policies to enhance adoption industry-wide • Use incentives like deposit/refund schemes to support circular practices in construction. • If on-site recycling is mandated, the policy could better support circularity in the sector.
2	Effective demolition planning	Proper demolition planning can improve material supply and waste segregation for recycling and reuse in the construction sector	11	Streamlined policy procedures	• Streamline deposit/refund systems to reduce barriers and encourage responsible waste management. • A simplified approach could encourage recycling in smaller projects without imposing too many procedural burdens.
3	Immediate supply availability	Enabling recycling practices ensures a ready supply of materials, minimising potential project delays	12	Enforcement of large organisations	Implement deposit schemes first for larger companies to ease adoption and reduce pressure on smaller enterprises.
4	Transportation and accessibility	The availability of accessible recycling facilities and proper transportation options supports material recycling efforts	13	Prevention of stockpiling	Address potential stockpiling through complementary measures and supply chain support.
5	Circularity model requirement	Establish effective circularity models to meet recycling and disposal demand in the construction sector	14	Refund amount	Ensure sufficient refund amounts to encourage effective waste management and support circularity in construction.
6	Recycling infrastructure development	Develop new recycling facilities to meet growing recycling demands	15	Adaptation for market and policy needs	Align recycling incentives with local and federal policies to encourage innovation and adaptability in recycling.

Table 18. (Continued)

No	Measure	Description	No	Measure	Description
7	Policy impact on supply chain	Highlight that diverting material from landfills can help build recycling infrastructure and improve supply chain efficiency	16	Addressing asbestos concerns	Market development and asbestos contamination concerns must be addressed for safe material reuse.
8	Partnership with the waste industry	• Collaborate with industry to encourage recycling, efficient demolition, and PwRC use • Partnering recycling businesses with demolition projects can reduce waste management costs and improve waste diversion outcomes	17	Financial incentives for waste management	• Provide financial incentives like refunds to offset waste management costs and divert waste from landfills. • Incentives could foster a new deconstruction market, supporting circular economy efforts in the building sector.
9	Tracking system	• Develop material tracking systems and offer higher returns for valuable materials to promote waste management • Implement consistent tracking systems from demolition to recycling for quality control and environmental impact management	18	Applicability challenges	Address challenges like brand ownership and long-term deposit management in buildings with extended lifespans.

Source: Authors.

Table 19. Key measures to enhance the effectiveness of the proximity principle policy

No	Measure	Description	No	Measure	Description
1	Regional recycling hubs	• Develop community hubs, like council tip stores, to facilitate material exchange and promote local reuse and circular economy practices • Facilitate material exchange and encourage circular practices through community recycling hubs	9	Government mandates	• Mandate local sourcing for government projects to drive demand for local PwRC. • Prioritise local recycling for public projects to boost the local economy and reduce the carbon footprint. • Enact policies promoting proximity-based waste management and local PwRC usage to strengthen sustainable practices.

No	Measure	Description	No	Measure	Description
2	On-Site recycling	• Encourage on-site recycling to reduce transportation needs and enhance waste processing efficiency on the construction site	**10**	Online material exchange	• Use online platforms for material exchange to support local reuse and circularity in construction projects.
3	Addressing regional adaptation challenges	• Address challenges in implementing proximity practices in rural or geographically dispersed areas • Support infrastructure that enables sustainable local disposal and recycling, particularly in remote areas	**11**	Balancing supply and demand	• Align supply and demand for PwRC in local sourcing to improve sustainability without reducing reusability efficiency. • Partner with local recycling centres to integrate PwRC into construction supply chains, reducing reliance on virgin resources • Source locally to reduce transportation and support emissions reduction targets
4	Incentives for local sourcing	Offer financial incentives for local sourcing to support the economy and lower emissions from long-distance transport	**12**	Standardised reporting	Use standardised reporting to monitor local emissions savings and improve compliance and performance tracking.
5	Industry education & public awareness campaigns	• Providing education on the benefits of PwRC and proper recycling techniques fosters support for local waste management • Educate stakeholders on PwRC benefits to build broader support and demand for sustainable construction practices • Providing education to contractors on local waste processing and recycling practices promotes compliance with proximity principles	13	Defined policy radius	• Specify a policy radius based on proximity to recycling facilities to prioritise local processing. • Define clear boundaries for waste processing policies to align with environmental goals and support local suppliers. • Engage recyclers in defining boundary areas to ensure policies meet industry needs effectively.
6	Additional frameworks required	Implement additional frameworks to support circularity in the building and construction sector	14	National harmonised application	Expand proximity initiatives nationally to enhance waste reduction efforts

Table 19. (Continued)

No	Measure	Description	No	Measure	Description
7	Impact on early design decisions	Integrate waste diversion targets into early project designs to reduce transport needs and support circular economy objectives	15	Development of eco-parks	• Foster circular material flows through industrial hubs or eco-parks • Promote local processing and reuse of PwRC through eco-parks and industrial symbiosis
8	Availability of recycling options	• Ensure sufficient recycling facilities are near construction sites for effective local waste management. • Promotes local recycling centres near construction sites, reducing waste transportation and increasing local PwRC adoption • Recognise that insufficient local facilities can increase transportation distances, limiting proximity policy impact.			

Source: Authors.

Table 20. Key measures to enhance the effectiveness of banning the export of C&D waste resources

No	Measure	Description	No	Measure	Description
1	Application of landfill levy	• Generate stricter policies on exporting C&D waste and raising landfill levies will foster PwRC and sustainable development domestically	7	Certification and labelling	• Foster confidence in domestic markets and drive demand for PwRC in construction projects through the utilisation of transparent certification and labelling systems
2	Market development	• Creating long-term demand for PwRC and investing in market development helps establish a sustainable local supply chain for PwRC	8	Regulatory support	• Develop supportive waste regulations and clear quality standards for PwRC to mainstream their use in the construction sector.

No	Measure	Description	No	Measure	Description
3	Investment in domestic infrastructure	• Invest in recycling centres and processing facilities to support local C&D waste management and reduce export reliance • Enhance domestic recycling infrastructure to avoid landfill disposal • Pair export restrictions with infrastructure investment to prevent illegal dumping • Plan for increased investments in recycling plants and local processing facilities • Provide financial assistance for recycling start-ups to enhance local processing capabilities, meeting the demand generated by export bans • Balance regulatory enforcement and infrastructure capacity to manage increased recycling demands domestically	9	Effective policy enforcement	• Apply strict enforcement and inspections to ensure compliance with PwRC policies and prevent illegal exports. • Provide strong regulatory support and penalties for non-compliance to ensure the effectiveness of waste export restrictions and drive companies toward sustainable practices. • Enforce strict standards on handling contaminated materials domestically to ensure better control and quality in recycling processes.
4	Sustainable procurement	Consider subsidies and procurement preferences for PwRC in public projects to reduce export reliance and promote domestic recycling	10	Innovation and technology	Invest in R&D for advanced C&D waste recycling technologies to reduce export reliance.
5	Collaboration with industry	• Engage stakeholders in joint efforts to increase awareness and uptake of PwRC in construction projects to encourage recycling and support market uptake of recycled materials domestically • Build capacity to process and recycle C&D waste locally, supporting circular economy and reducing environmental impacts	11	Priority material analysis	• Analyse materials to determine domestic recyclability versus those requiring export • Identify high-value materials that can be effectively processed and recycled locally to maximise the utility of restricted materials and align with the circular economy goals. • Analysis of onshore processing feasibility for specific materials like asbestos, requiring appropriate infrastructure
6	Education and outreach	• Conduct outreach campaigns to inform stakeholders of the environmental benefits of PwRC to encourage more responsible practices and support domestic recycling efforts. • Raise awareness of local C&D processing benefits to reduce exports and encourage more support for domestic recycling policies. • Highlight those restricting exports to reduce emissions from transport, support the carbon goals of circular economy policies and favour domestic processing.			

Source: Authors.

Table 21. Key measures to enhance the effectiveness of product stewardship policy

No	Measure	Description	No	Measure	Description
1	Support local recyclers	Support local recyclers in product stewardship programs to strengthen circular economy practices and mitigate overseas waste dependency	9	Government role and enforcement	• Government enforcement ensures compliance with product stewardship policies, preventing rogue operations. • Government support, including grants and policies, fosters industry compliance and investment in sustainable practices
2	Circular design approach	• Encourage designers to choose recyclable, easily dismantled materials, reducing environmental impact over a product's lifecycle • Implement design for recycling to simplify recycling and resource recovery, reducing costs and environmental impacts • Have a lifecycle focus to encourage sustainable design choices, enhancing PwRC utilisation from the initial stages	10	Analyse and manage implementation challenges	• Understand and address implementation challenges like high collection costs and material recovery across diverse regions. • Lack of control over end-of-life for certain products limits recycling efficiency without supportive policies.
3	Role of co-regulators	Appointing co-regulators, such as local recycling organisations, can help manage and support product stewardship policies effectively	11	Standardisation of stewardship programs	Establishing clear, uniform standards for product stewardship can address quality inconsistencies and improve outcomes.
4	Education and awareness	• Raise awareness about product stewardship to help suppliers and manufacturers adopt sustainable practices • Educate suppliers, manufacturers, and contractors on product stewardship to facilitate mainstreamed PwRC adoption • Highlight that product stewardship programs assign responsibility to manufacturers to promote recycled content and infrastructure investment	12	Funding and financial support	• Financial aid and incentives for SMEs to enhance product stewardship efforts and encourage sustainable design practices • Provide tax credits, grants, and support for PwRC use to enable sustainable practices in the construction sector. • Support recycling start-ups and provide funding to enhance the local recycling ecosystem and achieve stewardship objectives.

No	Measure	Description	No	Measure	Description
5	Effective reporting	Require product stewardship schemes to publicly report recycled product volumes to enforce environmental responsibility	13	Managing product waste responsibility locally	Supporting local recyclers strengthens circular economy practices and reduces overseas waste dependency.
6	Policy design	• Develop minimum standards to ensure product stewardship policies effectively meet waste management goals • Draw inspiration from tyre stewardship, which highlights the potential benefits and challenges of structured recycling systems. It underscores the importance of stricter regulations to prevent free riders and ensure the system's effectiveness • Mandated stewardship schemes achieve higher compliance than voluntary programs, which may result in limited participation • Account for true recycling costs to make PwRC economically feasible	14	Certification and labelling	• Utilise certification schemes and product labels to guide consumers toward PwRC • Apply environmental impact labelling that details recycled content percentages to drive informed choices among builders and developers. • Clear labelling and lifecycle tracking of products build trust and traceability, encouraging the adoption of PwRC. • Utilise chain of custody standards, like FSC or Programme for the Endorsement of Forest Certification (PEFC) for timber to enhance material transparency and full lifecycle accountability
7	Investment in recycling infrastructure	• Development of waste infrastructure support enables the handling and recycling of construction waste, preventing overseas export and promoting local recycling • Sustained support and investment are essential for significant waste reduction and recycling reliability	15	Supply chain collaboration	• Make cross-supply chain efforts with suppliers and contractors to drive PwRC use and streamline recycling processes, benefiting the building and construction sector. • Collaborate with all parties in the supply chain to create continuity for PwRC use and ensure end-of-life management.
8	Conducting LCA	Conduct lifecycle assessments to inform environmentally sound construction material choices.			

Source: Authors.

Table 22. Key measures to enhance the effectiveness of levy waiver waste recycling residual

No	Measure	Description	No	Measure	Description
1	Auditing and verification	• Conduct audits to prevent misuse of waivers and ensure adherence to recycling practices • Implement verification systems to maintain program integrity and reduce loopholes • Ensure transparency in waiver audits to build trust and hold stakeholders accountable for meeting their recycling targets • Introduce voluntary self-audits while ensuring safeguards against misuse	9	Clarity in the policy criteria and definitions	• Provide transparent criteria for waivers to manage residual waste responsibly • Provide clear and specific definitions are needed for recycling waivers to prevent loopholes Ensure waiver policies are specific and clear to avoid exploitation and ensure compliance • Address vagueness in current waiver policies to prevent "greenwashing" by some organisations • Establishing clear criteria for waivers incentivises continuous improvement (i.e. performance-based criteria) and defines the requirements for recycling eligibility • Waivers must ensure that organisations meet set quality standards to expedite processing while maintaining adequate oversight and quality control
2	Financial incentives	Provide tax breaks and grants to entities that exceed PwRC targets	10	Collaboration and partnerships	Foster partnerships among government, industry, and community groups to maximise recycling efforts
3	Potential adverse impacts	Recognise and address potential negative impacts of the policy on PwRC usage	11	Adjust pricing	• Adjust pricing based on the recyclable portion of waste to ensure organisations are not overcharged for residual waste disposal • Adjusting waste drop-off prices based on recyclability could incentivise better recycling practices
4	PwRC utilisation minimum	Link waiver eligibility to meeting PwRC usage to encourage organisations to meet recycling goals and support sustainable practices in the construction sector	12	Material innovation support	• Expand waivers to include innovative materials that may not currently be recyclable but have potential in future sustainability practices
5	Documentation and reporting	Require regular recycling documentation to promote transparency and accountability	13	Alignment with other policies	Should work in tandem with other policies for effective implementation

No	Measure	Description	No	Measure	Description
6	Investment in recycling infrastructure	Waivers should encourage entities to invest in recycling technology, funding upgrades, or infrastructure to boost recycling outcomes	14	Policy regular evaluation	Conduct regular program evaluation to ensure that the waiver remains aligned with current recycling objectives and adjusts to feedback
7	Recognition and incentives	• Recognise organisations that exceed recycling targets with awards and certifications • Organisations delivering tangible recycling outcomes could benefit from reduced levies, promoting higher commitment to sustainability	15	Waivers prioritisation	• Prioritise waivers for materials with limited recycling options (hard-to-recycle materials) to encourage innovation • Prioritise waivers for organisations using local PwRC sources to strengthen regional recycling networks • Allow waivers for entities with strong recycling commitments to support responsible waste management while avoiding unnecessary fees
8	Education	Support initiatives to raise public and internal awareness about recycling's importance			

Source: Authors.

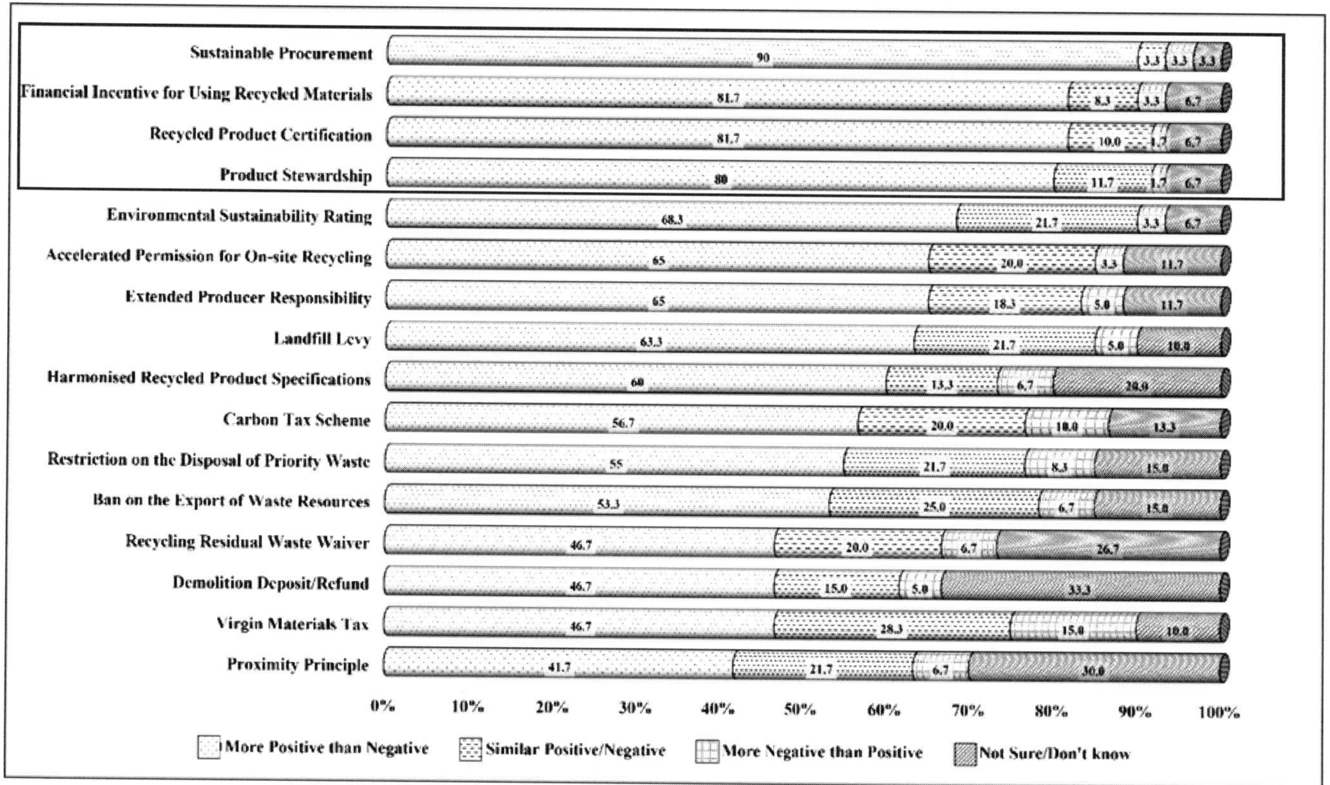

Source: Authors.

Figure 5. Participants' perspectives on the positive vs negative impacts of each CE policies.

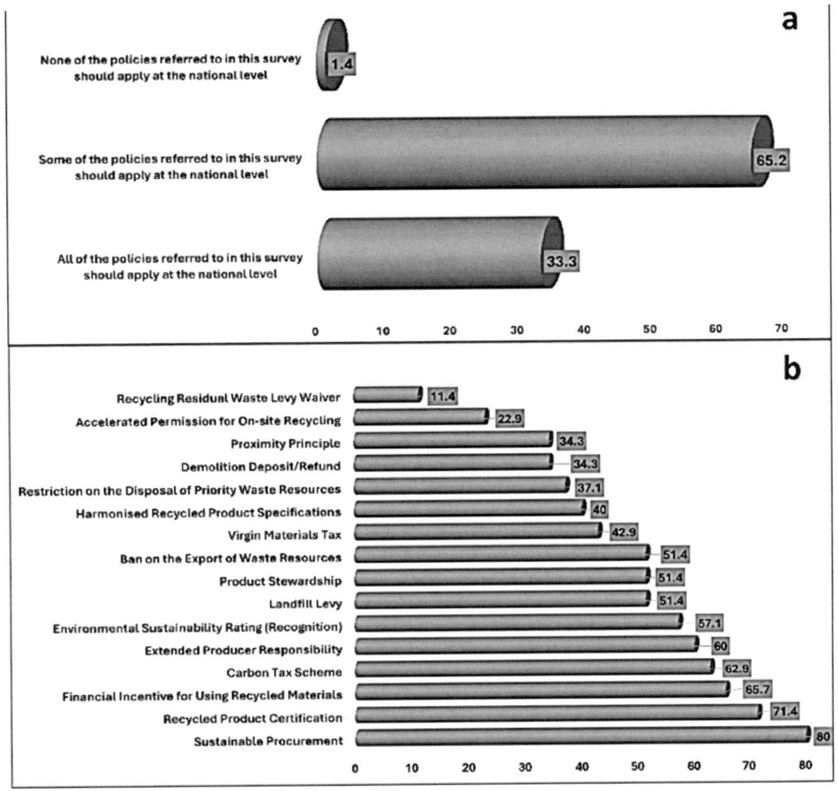

Source: Authors.

Figure 6. Participants' perspectives on the implementation scale of each CE policy (frequency distribution %).

4.3.2. Implementation Strategies and Key Measures

The participants were asked to provide recommendations to enhance the optimal utilisation of PwRC in construction projects and if they have any strategies to address the issues related to each policy implementation. Not all respondents provided recommendations for the study policies. For better analysis, comments with similar nature and direction were grouped. The following series of tables (Tables 6-22) provides a summary of key measures that can be taken to adopt each policy for enhancing the use of PwRC in the construction sector.

5. Comparative Analysis and Further Feedback

Based on all the findings of the survey detailed in the previous section of the chapter, the following provides the comparative analysis and the further feedback that was provided by the respondents that can help to enhance the utilisation of PwRC in the construction sector. Benefits, challenges and impacts of policies

The participant's responses on the impact mechanism for each policy were comparatively analysed to identify the five most pivotal policies that have the greatest impact on each impact mechanism. The following bullet points outline the top five policies identified for each impact mechanism:

- *Building end-user confidence in PwRC:* The five most impactful policies are EPR, Environmental Sustainability Rating, Harmonised PwRC Specifications, Recycled Product Certification, and Sustainable Procurement.
- *Creating demand for PwRC:* The top policies driving demand include Virgin Material Tax, EPR, Financial Incentives, Carbon Tax Scheme, and Sustainable Procurement.
- *Establishing sustainable PwRC supply chains:* Key policies for effective supply chain creation are EPR, Sustainable Procurement, a Ban on the Export of C&D Waste, Financial Incentives, and Virgin Material Tax.
- *Enhancing environmental sustainability in construction:* The leading policies for sustainability improvement are the Proximity Principle, EPR, Accelerated On-Site Recycling Permissions, Restrictions on the Disposal of Priority Waste, and the Carbon Tax Scheme

- *Improving PwRC quality:* The main policies influencing quality include EPR, Harmonised PwRC Specifications, Sustainable Procurement, Recycled Product Certification, and the Carbon Tax Scheme.
- *Boosting the social image of PwRC end-user organisations:* Key policies enhancing social perception include EPR, Environmental Sustainability Rating, Accelerated On-Site Recycling Permissions, Sustainable Procurement, and Recycled Product Certification
- *Enhancing working conditions with PwRC:* The most effective policies for improved working conditions are Environmental Sustainability Rating, Accelerated On-Site Recycling Permissions, Harmonised PwRC Specifications, Restrictions on the Disposal of Priority Waste, and Demolition Deposit Refund
- *Influencing decision-making on PwRC use:* The primary policies impacting decision-making are the Environmental Sustainability Rating, EPR, Carbon Tax Scheme, Sustainable Procurement, and Demolition Deposit Refund.
- *Ensuring a sustainable PwRC supply:* Policies supporting a reliable PwRC supply include EPR, Demolition Deposit-Refund, Ban on the Export of C&D Waste, Restrictions on the Disposal of Priority Waste, and Accelerated On-Site Recycling Permissions
- *Reducing PwRC costs:* Top policies for cost reduction include Accelerated On-Site Recycling Permissions, Financial Incentives for PwRC use, the Proximity Principle, Carbon Tax Scheme, and EPR.

5.1. Further Policies and Participants' Descriptive Feedback

At the end of the survey, participants were invited to share their knowledge of any additional policies not covered in this study. Among the respondents, 22 indicated awareness of supplementary policies or related directives, while 39 reported no knowledge of policies beyond those addressed in the survey. Table 23 provides a summary of participants' responses on this topic.

Table 23. Summary of participant's perspective regarding additional circular economy policies

Category	Description
Implementation & Practice	The side system needs better implementation, traceability, quality, and industry-standard practices.
Specific Policies Mentioned	• NSW Sustainability Policy, Circular Economy Action Plan, LCA targets, Recycled First Policy, State Environmental Planning Policies, EPA requirements, Circular Economy in NSW, Digital • Product Passports, Right to Repair, Traceability Standards, and Federal Social and Environmental Product Procurement
Policy Suggestions or Gaps Identified	Suggestions include waste reduction incentives, tax benefits for households and companies, waste credits, mandatory traceability and reporting, investing profits from mining and gas into sustainability, affordable housing initiatives, promoting C&D waste for private sector usage, and incentives for circular economy in the private sector.
Noteworthy Examples Cited	Policies in EU countries (e.g., Sweden's construction waste guidelines), Recycled First in Victoria, London's Circular Economy Statement, NSW Circular Design Guide, Modernisation Fund for waste recycling.
Feedback on Current Policy Landscape	Some find existing policies one-dimensional, suggesting that they take a multi-dimensional approach. Respondents mention the need for alignment between policy and infrastructure roadmaps, as well as the inclusion of conditions in business leases to boost PwRC demand.
Awareness Challenges	Low awareness in the sustainability industry despite time in the field; limited school education on sustainability; greater familiarity with circular economy policies and use of secondary raw materials seen as a need

Participants also shared their perspectives on the implementation of circular economy policies designed to enhance the utilisation of PwRC in construction projects. Comments emphasised the need for better enforcement of policies, along with actionable outcomes. Advocacy for national standards was also supported, with some stating policies should be complementary and not conflicting to maximise impact. Feedback highlighted that this field is new in Australia, with limited familiarity for many in industry or government. The complexity due to Australia's geographic and demographic context was also mentioned, along with limited awareness of policies across respondents.

Conclusion

This chapter explored the perceptions of Australian stakeholders concerning CE policies aimed at promoting the use of PwRC within construction projects. The findings reveal varied levels of familiarity and nuanced views on the effectiveness of different CE policies, highlighting certain limitations and potential measures for improving their implementation. Responses regarding whether policies should be applied nationally or at the state level underscore the complexity of aligning CE approaches across Australia's diverse regulatory landscapes. Specifically, while some stakeholders advocate for a national framework to ensure uniformity and scalability, others argue that state-specific policies are better suited to address local resource recovery needs and infrastructure constraints.

The research findings make significant contributions to knowledge development by mapping current awareness gaps and perceived barriers to the application of CE policies. These insights are essential for academics and policymakers seeking to understand the obstacles to CE adoption and identify areas where targeted education or policy adjustments are necessary. The findings also serve as empirical evidence for policy development, pinpointing the specific areas where policy clarity or standardisation is needed to boost stakeholder confidence and engagement in circular practices and the optimal utilisation of PwRC. For industry practice, the research provides actionable insights, particularly in identifying practical measures that stakeholders believe could enhance CE policy effectiveness, such as improved incentives, clearer communication of policies, and alignment of CE objectives across construction project phases.

For the building and construction and waste resource sectors, the research offers insights to refine current practices. Construction organisations can leverage these findings to integrate CE principles more effectively into project planning and material procurement, focusing on materials with established recycling pathways and minimal environmental impact. Waste management and resource recovery organisations may use these insights to align processing capabilities with the demand generated by new CE policies, supporting the industry's transition towards higher recycled content usage. Industry peak bodies have a crucial role in disseminating these findings to promote the circular economy. By developing guidelines, training programmes, and policy recommendations informed by this research, peak bodies can encourage the optimal use of PwRC and enhance collaboration between stakeholders, driving industry-wide progress towards sustainable resource management.

Overall, this chapter provides policymakers, industry leaders, and peak bodies with a strong understanding of current perceptions and actionable pathways to enhance CE policy efficacy, supporting a sustainable transition for Australia's construction and waste management sectors.

References

[1] Shooshtarian, S., Maqsood, T., Wong, P., Khalfan, M. and Yang, R.J., Market development for construction and demolition waste stream in Australia. *Journal of Construction Engineering, Management & Innovation*, 2020. 3(3): p. 220-231.

[2] Stijn, A.v. and Gruis, V. *Towards a Circular Economy in the Built Environment: An Integral Design Framework for Circular Building Components*. 2020.

[3] Hossain, M.U., Ng, S.T., Antwi-Afari, P. and Amor, B., Circular economy and the construction industry: Existing trends, challenges and prospective framework for sustainable construction. *Renewable and Sustainable Energy Reviews*, 2020. 130: p. 109948.

[4] Osobajo, O.A., Oke, A., Omotayo, T. and Obi, L.I., A systematic review of circular economy research in the construction industry. *Smart and Sustainable Built Environment*, 2022. 11(1): p. 39-64.

[5] Iyer-Raniga, U., Gajanayake, A. and Ho, O.T.-K., The Transition to a Circular Built Environment in Australia: An Analysis of the Jurisdictional Policy Framework. *Environmental Policy and Law*, 2023(Preprint): p. 1-14.

[6] Shooshtarian, S., Caldera, S., Maqsood, T. and Ryley, T., Using recycled construction and demolition waste products: A review of stakeholders' perceptions, decisions, and motivations. *Recycling*, 2020. 5(4): p. 31.

[7] Klein, N., Deutz, P. and Ramos, T.B., A survey of Circular Economy initiatives in Portuguese central public sector organisations: National outlook for implementation. *Journal of environmental management*, 2022. 314: p. 114982.

[8] Ren, Q. and Albrecht, J., Toward circular economy: The impact of policy instruments on circular economy innovation for European small medium enterprises. *Ecological Economics*, 2023. 207: p. 107761.

[9] Shooshtarian, S., Hosseini, M.R., Kocaturk, T., Arnel, T. and T. Garofano, N., Circular economy in the Australian AEC industry: investigation of barriers and enablers. *Building Research & Information*, 2023. 51(1): p. 56-68.

[10] Saez, P.V., del Río Merino, M., González, A.S.-A. and Porras-Amores, C., Best practice measures assessment for construction and demolition waste management in building constructions. *Resources, Conservation and Recycling*, 2013. 75: p. 52-62.

[11] Salkind, N.J., *Encyclopedia of research design*. Vol. 1. 2010: sage.

[12] Keena, N., Rondinel-Oviedo, D.R. and Demaël, H. *Circular Economy Design towards Zero Waste: Laying the foundation for constructive stakeholder engagement on improving construction, renovation, and demolition (CRD) waste management*. in *IOP Conference Series: Earth and Environmental Science*. 2022. IOP Publishing.

[13] Vasconcelos, L.T., Silva, F.Z., Ferreira, F.G., Martinho, G., Pires, A. and Ferreira, J.C., Collaborative process design for waste management: co-constructing strategies with stakeholders. *Environment, Development and Sustainability*, 2022. 24(7): p. 9243-9259.

[14] Frempong-Jnr, E.Y., Ametepey, S.O. and Cobbina, J.E., Impact of stakeholder management on efficient construction waste management. *Smart and Sustainable Built Environment*, 2023. 12(3): p. 607-634.

Index

A

B

C

D

E

S

T

U

V

W